About IFPRI

The International Food Policy Research Institute (IFPRI®) was established in 1975 to identify and analyze alternative national and international strategies and policies for meeting food needs of the developing world on a sustainable basis, with particular emphasis on low-income countries and on the poorer groups in those countries. While the research effort is geared to the precise objective of contributing to the reduction of hunger and malnutrition, the factors involved are many and wide-ranging, requiring analysis of underlying processes and extending beyond a narrowly defined food sector. The Institute's research program reflects worldwide collaboration with governments and private and public institutions interested in increasing food production and improving the equity of its distribution. Research results are disseminated to policymakers, opinion formers, administrators, policy analysts, researchers, and others concerned with national and international food and agricultural policy.

About IFPRI Research Reports

IFPRI Research Reports are well-focused, policy-relevant monographs based on original and innovative research conducted at IFPRI. All manuscripts submitted for publication as IFPRI Research Reports undergo extensive external and internal reviews. Prior to submission to the Publications Review Committee, each manuscript is circulated informally among the author's colleagues. Upon submission to the Committee, the manuscript is reviewed by an IFPRI reviewer and presented in a formal seminar. Three additional reviewers—at least two external to IFPRI and one from the Committee—are selected to review the manuscript. Reviewers are chosen for their familiarity with the country setting. The Committee provides the author its reaction to the reviewers' comments. After revising as necessary, the author resubmits the manuscript to the Committee with a written response to the reviewers' and Committee's comments. The Committee then makes its recommendations on publication of the manuscript to the Director General of IFPRI. With the Director General's approval, the manuscript becomes part of the IFPRI Research Report series, which began in 1977.

Publications Review Committee 2009

About IFPRI

About IFPRI Research Reports

Successful Organizational Learning in the Management of Agricultural Research and Innovation

The Mexican Produce Foundations

Javier M. Ekboir, Gabriela Dutrénit, Griselda Martínez V.,
Arturo Torres Vargas, and Alexandre O. Vera-Cruz

RESEARCH REPORT 162

IFPRI®

INTERNATIONAL FOOD POLICY
RESEARCH INSTITUTE

sustainable solutions for ending hunger and poverty

Supported by the CGIAR

International Food Policy Research Institute
2033 K Street, NW
Washington, D.C. 20006-1002, U.S.A.
Telephone +1-202-862-5600
www.ifpri.org

DOI: 10.2499/9780896291713RR162

Library of Congress Cataloging-in-Publication Data

Successful organizational learning in the management of agricultural research and innovation : the Mexican produce foundations / Javier M. Ekboir . . . [et al.].
p. cm.—(IFPRI research report ; 162)
Includes bibliographical references.
ISBN 978-0-89629-171-3 (alk. paper)
1. Agriculture—Research—Mexico—Finance. 2. Agricultural innovations—Mexico—Finance. 3. Organizational learning. I. Ekboir, Javier. II. International Food Policy Research Institute. III. Title: Mexican produce foundations. IV. Series: Research report (International Food Policy Research Institute) ; 162.
S542.M6S83 2009
338.1′30972—dc22 2009004766

Cover photo: © IFAD/Franco Mattioli

Contents

Tables

Figures

Boxes

Foreword

Traditional innovation and extension systems seem to have become less effective, and new approaches and instruments are needed. To develop these instruments, all actors in agricultural innovation systems have to build up their innovation capabilities. The analysis of these capabilities, both in individuals and organizations, has received a great deal of attention outside agriculture. Innovation capabilities within agriculture, however, have barely been explored. Equally unexplored is the issue of how to strengthen innovation capabilities in nonprofit organizations that seek to alleviate poverty. This study contributes to these hitherto underdeveloped research areas by building a novel conceptual framework to analyze the dynamics of innovation capabilities in an organization that supports agricultural research, extension, and innovation in Mexico. Derived from a case study, the report's conclusions are relevant to a wide range of organizations both in developed and developing countries.

The results show that an organization's innovation capabilities depend on the interactions among committed, capable individuals; on the organization's culture and governance mechanisms; and on the enabling environment, all of which are conditioned by the organization's history. Building innovation capabilities is a complex process, and there are no simple formulas for accomplishing it; all successful organizations will face difficulties when adapting to a new situation. However, the authors conclude that three factors are essential in helping organizations to innovate: (1) they must have in their ranks at least a few committed, inventive individuals; (2) those individuals must be given latitude, within the framework of the organization's goals, to experiment with new or different approaches; and (3) funding institutions (whether international donors or public offices) have to allow innovative organizations to explore new instruments to achieve the institutional goals.

The people and leadership factors seem to remain undervalued and understudied as key forces for innovation and agricultural development in general.

Joachim von Braun
Director General, IFPRI

Acknowledgments

Many people helped us during this research project, and it is impossible to mention them all. We especially thank Carlos Baranzini Coronado, president of the National Coordinator for the Produce Foundations (COFUPRO), and Raul Romo, COFUPRO's executive secretary, for their support and help during the many years we worked with the Produce Foundations. We also thank Marta Xochitl Flores Estrada, president of the Produce Foundation of Michoacán, for encouraging us to think not only as researchers but also as partners of the many actors in the Mexican agricultural innovation system. The recent experience of implementing an action research program to build organizational capabilities for agricultural innovation has been extremely enlightening for us, well beyond what we could have learned with traditional research methods. We also thank all the interviewees for their willingness to spare their time (sometimes several hours of it) and share their experiences. We thank the Publications Review Committee and two anonymous reviewers for insightful comments that helped us greatly improve the original manuscript and transcend our Mexican perspective; the result is a report that we hope will be useful to broad audiences both in developed and developing countries. We acknowledge COFUPRO's and the International Food Policy Research Institute's (IFPRI) financial support for this project and the National Council for Science and Technology's (CONACYT) support for the second stage of the research program. Finally, we thank Isabel Rodríguez for her research assistance and John Whitehead for his editorial support.

Financial Contributors and Partners

IFPRI's research and capacity-strengthening and communications activities are made possible by its financial contributors and partners. IFPRI receives its principal funding from governments, private foundations, and international and regional organizations, most of which are members of the Consultative Group on International Agricultural Research (CGIAR). IFPRI gratefully acknowledges the generous unrestricted funding from Australia, Canada, China, Finland, France, Germany, India, Ireland, Italy, Japan, Netherlands, Norway, South Africa, Sweden, Switzerland, United Kingdom, United States, and World Bank.

Acronyms and Abbreviations

ALNAP	Active Learning Network for Accountability and Performance in Humanitarian Action
CAS	complex adaptive system
CATIE	Centro Agronómico Tropical de Investigación y Enseñanaza (Tropical Agricultural Research and Higher Education Center)
CGIAR	Consultative Group on International Agricultural Research
CIAT	Centro Internacional de Agricultura Tropical (International Center for Tropical Agriculture)
CIP	Centro Internacional de la Papa (International Potato Center)
COFUPRO	Coordinadora Nacional de Fundaciones Produce (National Coordinator for the Produce Foundations)
CONACYT	Consejo Nacional de Ciencia y Tecnología (National Council for Science and Technology)
CORPOICA	Corporación Colombiana de Investigación Agropecuaria (Colombian Corporation for Agricultural Research)
CREA	Consorcios Regionales de Experimentación Agrícola (Regional Corsortia for Agricultural Experimentation)
CTN	Consorcio Técnico del Noroeste (Northwest Technical Consortium)
EMBRAPA	Empresa Brasileira de Pesquisa Agropecuária (Brazilian Agricultural Research Corporation)
FAO	Food and Agriculture Organization of the United Nations
FIRA	Fideicomisos Instituidos en Relación con la Agricultura, Banco de México (Agriculture-Related Trusts of the Central Bank of Mexico)
FMDR	Fundación Mexicana para el Desarrollo Rural (Mexican Foundation for Rural Development)
FOCIR	Fondo de Capitalización e Inversión del Sector Rural (Trust for the Capitalization and Investment of the Rural Sector)
GGAVATT	Grupos Ganaderos de Validación y Transferencia de Tecnología (Ranchers' Groups for Validation and Transfer of Technology)

ICAMEX	Instituto de Investigación y Capacitación Agropecuaria, Acuícola y Forestal del Estado de México (State Institute for Research and Training in Agriculture, Forestry, and Fishing)
IFAD	International Fund for Agricultural Development
IFPRI	International Food Policy Research Institute
IICA	Inter-American Institute for Cooperation on Agriculture
INIFAP	Instituto Nacional de Investigaciones Forestales, Agrícolas y Pecuarias (National Institute for Forestry, Agricultural, and Livestock Research)
ISNAR	International Service for National Agricultural Research
MERCOSUR	Mercado Común del Sur (Southern Common Market)
NGO	nongovernmental organization
PF	Fundación Produce (Produce Foundation)
PIEA	Patronato para la Investigación y Experimentación Agrícola (Foundation for Agricultural Research and Experimentation)
PRODESCA	Programa de Desarrollo de Capacidades en el Medio Rural (Program for Capacity Building in Rural Areas)
RFP	request for proposals
SAGARPA	Secretaría de Agricultura, Ganadería, Desarrollo Rural, Pesca y Alimentación (Secretary of Agriculture, Livestock, Rural Development, Fisheries, and Food)
SEDAGRO	Secretaría de Desarrollo Rural y Agricultura (Secretary of Rural Development and Agriculture)
SITT	Subprograma de Investigación y Transferencia de Tecnología (Subprogram for Research and Technology Transfer)
SNITT	Sistema Nacional de Investigación y Transferencia de Tecnología (National Research and Technology Transfer System)
STI	science, technology, and innovation
UAM	Universidad Autónoma Metropolitana (Autonomous Metropolitan University)
UNAM	Universidad Nacional Autónoma de México (National Autonomous University of Mexico)

Summary

Since the 1980s, developing countries' agriculture has become more complex and diversified. In general the public research and extension institutions in these countries were criticized for not participating in the emergence of the most dynamic agricultural markets. In recent years many of these institutions have struggled to adapt to the new environment, but they could not overcome the hurdles posed by organizational rigidities, strict public regulations, deteriorating human capital, shrinking budgets, and a model of science that hampered their integration into dynamic innovation processes.

In general developing countries applied similar agricultural research policies: separation of financing and implementation of research, reductions in direct budgetary allocations to research and extension institutions, elimination or major reduction of public extension, and introduction of competitive grants programs to induce a transformation of research organizations. Strong anecdotal information suggests that these policies had limited impact on the quality and pertinence of research and on the performance of the public research institutions.

Using a different set of instruments, the Mexican Produce Foundations (PFs) had major and diverse impacts on the agricultural innovation and research systems. These impacts resulted mostly from activities the PFs introduced as they learned to manage funds for research and extension, and to a lesser extent from the activities they were created for (that is, to manage funds for agricultural research and extension). The PFs were able to introduce these activities because they developed strong learning abilities, including identifying knowledge gaps and defining strategies to fill them.

The questions this report seeks to answer are how an organization that manages public funds for research and extension could sustain organizational innovations over extended periods and how it could learn and adapt to maximize its impact on the agricultural innovation system. Previous studies found that human resources, organizational cultures, and governance structures are three of the most important factors influencing institutional change and innovative capabilities. Despite their importance, these factors have been largely neglected in the literature on agricultural research and extension policies. This report analyzes what role these factors played in the Mexican experience.

What Are the PFs?

The PFs are civil society organizations managed by farmers. They were created to manage public funds for research and extension. The PFs were an institutional innovation of great importance in the Mexican innovation system and its agriculture. In the 2000s the PFs became an important stakeholder in the agricultural sector, influencing the design and implementation of agricultural policies, including scientific, technological, and innovation policies, as well as the transformation of public research institutions. New channels of communication were also opened between federal and state authorities on the one hand, and associations of commercial farmers on the other. These effects did not originate in the activities the PFs were created for but rather in activities the PFs themselves developed as they evolved.

The PFs were created at a critical moment in Mexico's history. In 1982 the country started a process of gradual economic and political liberalization. This process created new opportunities and threats for agricultural producers, who, in turn, started to look for advanced technologies to compete in the new economic environment. When public research institutions were not able to provide these technologies, producers and other actors in the innovation system imported or developed them.

It was in this context that 32 PFs (one in each state) were created to improve the interaction between the National Institute for Forestry, Agricultural, and Livestock Research (INIFAP) and farmers, as well as to boost its funding. Initially the federal and state governments controlled the boards of the PFs. A few farmers sought, and eventually achieved, a more independent role. Eventually this independence spread to all PFs. Subsequently the PFs realized that a set of isolated foundations was too atomized and created a national coordinating office (National Coordinator for the Produce Foundations, or COFUPRO).

Despite their greater independence, currently the PFs interact assiduously with the federal and state governments, and they influence sectoral and science policies. The PFs have also influenced the agricultural research system, especially by opening channels of communication between producers and researchers and by influencing research priorities. The PFs are a Mexican phenomenon in which foreign donors and multilateral institutions have had a very limited influence.

Conceptual Framework for the Analysis of Institutional Innovation

The PFs evolved in ways that were not totally intended or foreseen by any of the actors that participated in the process; even today, the impacts of their actions cannot be totally assessed. Such processes can be analyzed by combining complexity theories, an innovation systems framework, and the literature on organizational cultures and governance.

Complex Processes

The complex systems relevant to this study are made up of many interacting agents, each with his or her own goals and constraints. The set of agents self-organizes. This organization enables the emergence of structures and behaviors at higher levels of aggregation that are not possessed by any of the individual agents.

Complex systems evolve through the combination of initial conditions, multiple interactions, long-run trends, and random variations in the individual agents as well as in their interactions. Self-organization and randomness prevent individual agents from predicting or controlling the system's evolution. For this reason, policies in a complex system do not seek to direct the process but rather to raise the probability of desired outcomes while lowering that of undesired results, suggest new questions, examine new actions, and identify situations in which small interventions can have large impacts.

One of the most important instruments for operating on complex systems is to influence variation and selection. For example, a plant breeder makes crosses that do not occur naturally (an increase in variation). He or she then selects those plants with the best results according to his or her objective, regardless of reproductive efficiency (a change in the selection criterion). This mechanism differs fundamentally from the engineering approach to finding solutions. In complexity theory, many alternative solutions are tested to see which works, without fully understanding the underlying processes. In the engineering approach, a detailed model of the process is first created and then a solution is designed. The directed search converges to the optimum solution at least as quickly as the engineering solution; when the system is not easily characterized, a directed search is more efficient.

Organizational Innovative Capabilities: The Role of Learning

An organization's learning capacity stems from the interaction among resources (individuals and fixed capital), processes (how things are done), and values (including the organizational culture and mission). In new organizations, skills reside in resources, especially its human resources. In time, the skills in successful organizations are transferred to processes and values. Learning skills are specific to each organization and context—they are developed with idiosyncratic investments and processes that other organizations cannot easily imitate or buy. Skills that confer an advantage in one context may become a liability in another.

To be robust, a learning strategy should (1) include both operational and strategic components, (2) define what needs to be learned, (3) identify the sources of learning, (4) specify who learns and in which areas of the organization learning should occur, and (5) identify learning mechanisms.

Organizational Culture

An organizational culture is a set of basic assumptions discovered or developed by a group in the process of learning how to deal with problems of external adaptation and internal integration. These assumptions are considered to be valid because they have functioned adequately and are taught to new members of the organization. This definition highlights that a culture (1) is a matter of basic assumptions rather than values or behaviors; (2) is created by a group that devises, discovers, or develops these assumptions; (3) originates in the organization's initial needs to solve specific problems; and (4) is rooted in the mechanisms used to integrate workers. Cultures resist change because they generate great loyalty among their followers.

Governance

Organizations are currently recognized as spaces for collective action. Collective action depends as much on individual mental models as it does on organizational practices, such as the work environment and incentives offered to individuals. The analysis of governance in organizations has three dimensions: structure (distribution of functions and coordination), process (communication, coordination, leadership, learning policies, and operating processes), and strategic axis (mission, shared vision, strategic focus, and action plans).

Main Lessons Derived from the Study of the PFs

The study of the PFs yielded six important lessons for the management of research and innovation in developing countries. First, because of the complexity of innovation processes, science, extension, and innovation policies should be flexible and evolve as new information becomes available and new capabilities are acquired. The flexibility should be implemented at all levels: that of policymakers, research administrators, and researchers themselves.

Second, the development of innovative capabilities depends on a strong and sustained commitment from the authorities.

Third, research and innovation policies should not be left to evolve randomly but instead managed by balancing exploration of new instruments with the exploitation of those that have shown their efficacy. Balancing these two strategies requires an effective search mechanism that should combine decentralized exploration with centralized learning. Overreliance on centralized exploration can miss important opportunities, whereas decentralized learning alone hampers the sharing of useful information. Decentralized exploration can be complemented with directed searches when the opportunities or needs are clear. The directed searches should not attempt to establish major programs from the start but rather use pilot projects to test the assumptions underlying the program design. The pilot programs should be scaled up only after they have shown their effectiveness. Action research to test different types of interactions in the innovation networks should be the basic approach to pilot programs. The exploration should not be restricted to financing instruments and traditional research but should also include new ways to foster the emergence of innovation networks and involve new actors in innovation processes.

Another important element in the search strategy is a monitoring system to guide the exploration. The two most important factors that determine the system's efficacy are the flexibility of the monitoring system and the set of indicators to be monitored. A key feature of complex systems is that interventions often have unexpected results. A monitoring system that constrains itself to a predetermined set of indicators would miss those results.

Fourth, individuals play a major role in the success or failure of innovation processes and policies. Traditionally, policymakers have paid a great deal of attention to the design of the organizations that will implement the programs and the rules they must follow. The selection of capable individuals to run the organizations, however, has received less attention.

Fifth, innovation programs should have effective governance structures in which innovators can influence decisions.

Sixth, innovation is more than just applied research. To effectively participate in innovation processes, public research and extension institutions must adopt a new conceptual framework in which they recognize that they are not the central actors but play an important supporting role. Adoption of this new framework leads to new operational routines in which researchers integrate into innovation networks.

These lessons have particular relevance for poverty alleviation. Poverty and development are complex processes whose dynamics have changed in response to globalization, migration, and technical change. To operate successfully in such an environment, policymakers and practitioners need to develop effective exploration and selection mechanisms, create learning routines, and allow institutional change to foster adaptation to emergent problems. In addition, the institutions should be managed by individuals with a strong sense of duty and commitment to poverty alleviation.

As poverty alleviation was not among the original goals of the PFs, most of their actions were directed toward traditional research and extension activities. For this reason, the impacts of the PFs on rural poverty were never directly assessed. As they realized the limitations of traditional competitive funds, some PFs successfully implemented a few projects directed toward small farmers. Over the years, several PFs declared poverty alleviation an explicit institutional goal and devoted more resources to facilitate the access of small farmers to commercial value chains. There are also indications that the PFs contributed to poverty alleviation by supporting the expansion of commercial agriculture and fostering the creation of employment in rural areas. The impact of the PFs on research institutions and commercial farmers is currently being evaluated.

CHAPTER 1

Introduction

Since the 1980s, agriculture in developing countries has become more complex because of globalization, urbanization, the emergence of high-value agriculture, the deterioration of natural resources, migration, and changes in the livelihood strategies of rural households (Reardon 2005; World Bank 2006a, 2007).[1] These events have changed the dynamics of poverty, reducing the effectiveness of traditional approaches to poverty alleviation and creating the opportunity to use new instruments (see Section 3.4). Because of the complex nature of poverty, these instruments have to be designed using an evolutionary approach based on experimentation and learning (see Chapters 2 and 3).

After the Green Revolution, agricultural research was seen as an instrument in poverty-alleviation strategies. Since then, however, the perception of the role formal research can play has undergone shifts. Changes in perception were in part the consequence of the realization of the central role of growth in reducing poverty (World Bank 2005) and of the participation of different actors in the expansion of the most dynamic agricultural markets in developing countries. Private firms and service providers became important new sources of market and technical information, usually without government support, whereas the public research institutions continued to work on their traditional lines of research, especially staples, livestock, dairy, and a few traditional export products (Byerlee, Alex, and Echeverría 2002; see also the case studies in Weatherspoon and Reardon 2003; Gabre-Madhin and Haggblade 2004; and Pardey, Alston, and Piggott 2006).

The contrasting contributions of the private and public sectors led many stakeholders in the agricultural sector (including international donors) to question the usefulness of public research and extension institutions (Echeverría and Elliot 2002). This critique was part of a more general trend where policymakers began to expect greater direct usefulness from publicly funded research (Lundvall and Borrás 2005; Pavitt 2005).

Many public agricultural research and extension systems in developing countries have struggled to adapt to the new environment, but they could not overcome the hurdles posed by organizational rigidities, strict public regulations, deteriorating human capital, shrinking budgets, and a model of science that hampered their integration into dynamic innovation processes. Most of these factors also weakened policymakers' ability to transform the public institutions. In general, developing countries

[1]In this report the term *agricultural* refers to livestock, forestry, fishing, and agriculture itself. *Agriculture* is used for brevity.

applied a uniform recipe with little adaptation: separation of financing and implementation of research, reductions in direct budgetary allocations to research and extension institutions, elimination or major reduction of public extension, and introduction of competitive grants programs to induce a transformation of research organizations (see the cases described in Rivera and Alex 2004 and in Pardey, Alston, and Piggott 2006). Strong anecdotal information suggests that, in general, these policies had limited impact.

Using a different set of instruments, the PFs had major and diverse impacts on the agricultural innovation and research systems.[2] These effects resulted mostly from activities the PFs introduced as they learned to manage funds for research and extension, and to a lesser extent from the activities they were created for: to manage a competitive fund for agricultural research and extension.

The questions this report seeks to answer are how an organization that manages public funds for research and extension can sustain organizational innovation over extended periods, and how it can learn and adapt to maximize its impact on the agricultural innovation system. Studies of private firms have found that most organizations eventually lose their creativity and seldom regain it (see Chapter 3 for an analysis of the literature). The PFs, however, have managed to learn and adapt. Understanding the factors that enabled such unusual behavior will help to improve the design and implementation of research and innovation programs in developing countries.

Additional lessons can be obtained from new insights on the dynamics of innovative organizations and how they relate to innovative capabilities. Using a theoretical framework that combines the literature on innovation systems, complexity theories, and organizational cultures and governance, this study analyzes the factors that allowed the PFs to develop strong innovative capabilities and studies how these capabilities were affected by changes in the interactions among regulatory frameworks, organizational structures, creative individuals, and the history of the processes.

Previous studies found that human resources, organizational cultures, and governance structures are three of the most important factors influencing institutional change and innovative capabilities; moreover, it has been found that these factors evolve slowly, conditioned by the environment in which the organization operates (see Chapter 3). Despite their importance, these factors have been largely neglected in the literature on agricultural research and extension policies, which has mainly focused on estimating rates of return, developing formal methods for priority setting, and measuring inputs, especially public investments and numbers of professionals (for example, Alston, Norton, and Pardey 1995; Pardey, Alston, and Piggott 2006; and the Agricultural Science and Technology Indicators publications of the International Food Policy Research Institute [IFPRI]).

1.1 Why Study the PFs?

The PFs were an institutional innovation of great importance in the Mexican innovation system and its agriculture. Until the 1990s agricultural research and extension policies gravitated around INIFAP, but in the 2000s the PFs became one of the most important stakeholders in the agricultural sector and a de facto coordinator of several research and extension activities. The PFs influenced the design and implementation of agricultural policies, includ-

[2]The Spanish for PF is "Fundación Produce"; this term does not translate easily into English, meaning roughly, "Foundation Go Farm." The impacts were identified with rapid assessment methods in a previous evaluation of the PF (Ekboir 2004); a more thorough assessment of these impacts is currently under way.

ing science, technology, and innovation (STI) policies, as well as the transformation of public research institutions. New channels of communication were also opened between federal and state authorities on the one hand and associations of commercial farmers on the other. On the whole, these effects did not originate in the activities the PFs were created for (that is, to manage funds for agricultural research and extension), but rather in activities the PFs themselves developed as they evolved. Despite their focus on commercial farmers, some PFs have implemented a few innovative projects that targeted poor farmers; moreover, as time progressed, helping small farmers to access profitable markets and developing value chains that include poor households has become increasingly important for the most innovative PFs.

The PFs were created at a particular moment in Mexico's history. In 1982, the country suffered a major economic crisis, after which the government started a gradual economic and political liberalization.[3] The deregulation of the economy created new opportunities and threats for agricultural producers, who, in turn, started to look for advanced technologies to compete in the new economic environment. When public research institutions were unable to provide these technologies, producers and other actors in the innovation system imported or developed them (Ekboir et al. 2003). This lack of response from the public research and development institutions led society to question their validity.

It was in this context that 32 foundations (one in each state) were created to improve the interaction between INIFAP and farmers, as well as to boost funding for public research institutions. Initially the federal and state governments controlled the foundations' boards. Soon a few farmers sought, and eventually achieved, a more independent role. This independence gradually spread to most foundations. Before long the PFs realized that a set of isolated foundations was too atomized and created a national coordinating office (COFUPRO).

Despite their greater independence, the PFs interact assiduously with the federal and state governments. Most funding (82 percent) still comes from the federal government (Muñoz 2005), and the federal Secretary of Agriculture, Livestock, Rural Development, Fisheries, and Food (SAGARPA) consults frequently with the PFs about their activities and the transformation of the public research system. In adition, the PFs generate information and ideas that federal and state officials use to formulate agricultural policies, and the governments value the flexibility and administrative capacity of the PFs.

The PFs also influenced the agricultural research system. On the one hand, they introduced new concepts for the analysis and design of scientific and technological policies —especially the concepts of innovation and innovation networks. On the other hand, they also influenced research activities by opening channels of communication between producers and researchers and by defining new research priorities. The impacts on extension, though, do not appear to be substantial, mainly because there are no federal or state extension institutions with whom the PFs can interact.[4] In the face of this deficiency, several foundations created their own extension programs.

The PFs are a Mexican phenomenon. They were part of an international wave of creation of competitive funds for agricultural research and extension in developing countries, but beyond the intellectual input, foreign donors and multilateral institutions did not participate

[3]The nature of the crisis and how it influenced development policies is explained in Section 5.1.

[4]Public extension services were closed in 1996 and were replaced by a program intended to induce the creation of a market for technical services. Section 5.1 describes this process in detail.

in the creation of the PFs and contributed very little to their evolution.[5]

1.2 Motivation for This Study and Its Intended Audience

In 2005 the PFs thought they were losing their innovative drive and requested IFPRI to conduct an institutional assessment to identify actions to develop new capabilities. The PFs also thought that, as some of the founding members were leaving, it was necessary to document their experiences. The document submitted to the PFs is the basis of the present report and was modified to highlight those lessons that may be useful to an international audience of managers and funders of agricultural research and extension in developing countries (in particular, policymakers, research managers, public and private donors, and multilateral organizations) and the development specialists that advise them. These decision-makers are faced with the fact that traditional policies have lost much of their effectiveness in today's more complex rural environments. Thus they need new insights that help them (1) understand the role research and extension could play in growth and poverty alleviation in a globalized economy and (2) develop more effective instruments to design and implement research and innovation programs.

This report will also be useful for scholars who study development, innovation systems, research and extension policies, and management of innovative organizations, because it will help them identify issues for further research and new policy options that are both feasible and effective.

1.3 Outline of This Report

This report analyzes the institutional dynamics of a learning organization with the goal of helping policymakers in developing countries and multilateral organizations to improve the design and implementation of STI policies targeted at the agricultural sector in general, and at small, poor farmers in particular. For this reason, Chapters 2 and 3 contain a thorough review of a literature that is usually overlooked by economists and policymakers and can contribute to the identification of new instruments to foster rural innovations and improve poverty-alleviation programs. The main topics analyzed are complexity theories and the design of policies to operate on complex processes; the nature of innovation processes; STI policies; innovative capabilities; institutional learning; organizational cultures; governance structures; and how innovative capabilities influence poor households' livelihood strategies and their responses to poverty-alleviation programs.

Chapter 4 synthesizes the most important theoretical concepts used in the empirical analysis and describes the qualitative methodology used. Because qualitative methods are not commonly used by economists—and also because many social scientists and policy-makers believe that quantitative methods are always more reliable than qualitative approaches—this chapter also explains under which conditions one methodology should be chosen over another.

Chapter 5 describes in detail the creation and evolution of the PFs, in particular, the interaction among individuals, institutional settings, policies, the sociopolitical environment in which the PFs operated, and the process's own history. Three reasons justify the detail with which the process is described. First, it highlights the complex interactions among factors that are seldom considered in policy design and implementation; crucial among these factors are institutional cultures and the role of individuals and of overseeing institutions. Under-

[5]Similar programs have been identified in Argentina, Bolivia, Brazil, Chile, Ecuador, Nicaragua, Peru, the Central American region, and several African and Asian countries.

standing the roles of these factors and the complexity of socioeconomic processes yields policy recommendations substantially different from those derived from simpler, mechanistic analyses. Second, it shows how qualitative methods can be used to analyze the multiple and unintended impacts of socioeconomic policies. Finally, the report seeks to contribute to the creation of the institutional memory of the PFs.

Chapter 6 studies three important factors that influenced the evolution of the PFs: their institutional culture, governance structures, and learning mechanisms. These factors and their interactions are the main determinants of an organization's innovative potential, because they determine the exploration strategies and the ability to adapt and to influence the environment. This chapter also briefly explores some impacts the PFs had on the research and extension systems. Chapter 7 concludes and contains the main lessons learned from the analysis, and Chapter 8 describes the main actions the PFs implemented in response to the recommendations made in the original report finished in 2006.

The original report contained a detailed list of recommendations that could be implemented to strengthen the innovative capabilities of the PFs. Although many of the suggestions are quite specific to the PFs, they are included in Appendix C, because no other publication that detailed instruments for strengthening research managing institutions could be found in the literature.

CHAPTER 2

Innovation and Research Systems in Developing Countries

The PFs have greatly changed in their 10 years of existence. Today's PFs look very different from what their creators originally imagined: instead of being passive organizations subordinated to the political authorities, they are independent institutions often suggesting policies to a reactive public sector. The PFs evolved through the interactions of many actors, none of whom was able to control the process. The dynamics of such processes can be analyzed with complexity theories (Section 2.1). The characterization of innovation as a complex process—especially the relationship between STI and the role of networks in the generation and diffusion of innovations (Section 2.2)—helps identify new instruments to design and implement innovation policies. Finally, Section 2.3 analyzes STI policies from an innovation systems perspective.

2.1 Operating in a Complex World

2.1.1 What Is a Complex System?

The study of complexity theories is relatively recent (Nicolis and Prigogine 1989; Kauffman 1995, 2000; Axelrod and Cohen 1999; Watts 1999; Buchanan 2002; Crutchfield and Schuster 2003) and covers different areas of knowledge, including the social sciences, business management, physics, chemistry, ecology, biology, and mathematics.

There are different definitions of complexity; in this study, a complex system is defined as one whose properties cannot be understood from the separate analysis of its parts (Gallagher and Appenzeller 1999). There are also many types of complex systems. The most relevant for the analysis of organizational dynamics are made up of agents of different kinds (for example, directors, managers, employees, clients, and suppliers). Each agent defines his or her own strategy and reacts to environmental changes and to the actions taken by others; simultaneously, agents attempt to modify the environment to maximize their benefits. That is to say, the system is characterized by a number of independent decisionmakers, multiple interactions, and many feedback mechanisms operating at different scales (micro, meso, and macro) and speeds (that is, slow- and fast-changing variables interacting continuously). The nature and strength of the feedback mechanisms also change often. Such systems belong to a class known as complex adaptive systems (CASs). Examples of CASs are a colony of ants, large ecosystems, an army, and a market economy.

Patterns of behavior specific to groups of agents and to the system as a whole emerge from the actions and interactions among individual agents; many of the patterns do not exist at the level of the individuals. In other words, properties that manifest themselves at the meso or macro level do not exist at the micro level, but these properties influence micro behaviors. This phenomenon is known as self-organization (Kauffman 1995). For example, life results from a huge number of simultaneous chemical reactions. Age is a property of the body and not of any particular reaction; however, the reactions are influenced by age. Self-organization is one of the most important characteristics of a complex system. It is also the main reason that traditional methods of planning and policymaking fail when applied to a complex system (see Section 2.1.2).

CASs evolve through the combination of initial conditions, multiple interactions, trends, and random variations of both the agents and their interactions.[1] The relative importance of the trends and the random factors changes as the system evolves. When the trends are strong, the system is more or less predictable. With time, however, new agents, strategies, and interactions emerge, increasing instability. Eventually the random component predominates over the trends; at this point, known as a bifurcation, the system becomes completely unpredictable (Nicolis and Prigogine 1989).

CASs are characterized by many events occurring simultaneously; the majority of these events are of little consequence, a few have significant outcomes and, very rarely, catastrophic effects.[2] While trends predominate, the probability of the minor events occurring is greater, whereas close to the point of bifurcation, the probability of catastrophic events increases. Although it is possible to study the probability distribution of events, it is impossible to know whether the next event will have few consequences or will trigger a long chain of effects.[3] The study of past behaviors allows a better characterization of the probability distribution of events, but it does not eliminate uncertainty. For example, the movement of tectonic plates accumulates energy, which eventually causes an earthquake. When the accumulated energy is low, it is possible to predict with some certainty that an earthquake will not occur. As energy accumulates, the probability of an earthquake increases. But it is impossible to exactly predict either the timing or the intensity of the next quake.

At bifurcations, small events or agents with minimal influence can determine the future evolutionary path of a CAS, as seen in Box 2.1. The influence of minor individuals and/or events is greater at the beginning of the process because it is not yet structured. As time passes, organizations increasingly define routines and create their own cultures, which condition the actions of individuals (Christensen and Raynor 2003).

One of the most important factors in the evolution of a CAS is the interaction between variation and selection. New actors continually appear, but not all are well adapted to the environment. Selection allows for survival of the fittest. These concepts have traditionally been associated with the theories of Charles Darwin; nevertheless, new studies have shown that more complex dynamics than simple survival of the fittest are at play.

Studies of biological evolution have identified three characteristics of evolutionary processes. First, contrary to what Darwin believed,

[1]The initial conditions are those that prevail at the moment at which the analysis begins, which may differ from those present at the beginning of the process.

[2]In the mathematical literature, the term *catastrophic* refers to an event that essentially changes the structure of the process. It does not have the negative connotation that it has in common use.

[3]This phenomenon is known as critical self-organization (Bak 1996) or self-organization at the edge of chaos (Kauffman 1995).

> **Box 2.1 The influence of individuals and minor events on an organization's evolution**
>
> Bill Gates was contracted to develop an operating system for personal computers thanks to his mother's acquaintance with a top IBM executive. He was able to acquire a precursor of the operating system Windows from Xerox, because this company did not see its commercial value. IBM then used its marketing muscle to impose the standard, displacing other systems that were technically superior to Windows (Rycroft and Kash 1999). As almost all personal computers use Windows, there are few incentives for commercial software developers to write programs for other operating systems. At the same time, there is a great incentive for users of personal computers to choose Windows, as it allows them to easily exchange files with other computer users and ensures access to a continuous supply of compatible software. In other words, most personal computer users cannot choose alternatives that are more efficient than Windows because the cost of using other operating systems is too high.

the evolution of phenotypes does not indicate a movement toward an optimum.[4] Second, changes in biological efficiency do not occur gradually, but in leaps interrupted by long periods of relative stability. And third, genetic evolution, measured in terms of DNA sequences, does not stop during the period of phenotype stability but continues at least at the same rate as in the period of innovation and adaptation. The evolutionary leaps occur without external triggers, indicating the existence of "neutral" evolution; in other words, many genetic changes do not have a phenotypical expression.

It has been found that, except for fairly restricted settings, the evolution of complex systems cannot be characterized by the gradient of the equivalent of an energy function (Crutchfield and Schuster 2003). Thus for economics models, no welfare function (even a dynamic specification) can capture the rich dynamics of social systems.

The PFs can be analyzed as a CAS because:

1. They are part of and attempt to operate on a wider (innovation) system in which a large number of actors of different types (including various federal and state public offices, research and extension institutions, multilateral organizations, farmers, farmer organizations, and private companies) participate, and actors may have distinct objectives.
2. The diversity of the PFs, together with a lax system of governance, permits great variation in work styles, which converge through mechanisms of self-organization.
3. The actions of the PFs affect the innovation system through mechanisms that change in response to interventions of each of the system's actors, including the PFs themselves.

2.1.2 Policy Design in a CAS

After the industrial revolution, social processes have been analyzed using metaphors of machines and factories. This mechanistic vision had profound effects on organizational design, emphasizing control and predictability. In the public sector, priority was given to rules designed by hierarchies of relatively expert public employees and carried out by subordinates (Olson and Eoyang 2001).

In the framework of complexity theories, in contrast, self-organization is recognized both

[4]The genotype is the set of genes of an individual, whereas the phenotype is the visible expression of the genotype. For example, an individual can have genes for blue and brown eyes, but can only have eyes of one color.

as an essential characteristic that reflects the decentralized nature of the system and as the source of the difficulties in predicting the consequences of interventions. Self-organization prevents any individual agent from controlling the system; some actors have greater influence than others, but none is in complete control. For this reason, in complex systems decision-makers do not seek to direct the system but rather to operate on the probability of events to increase the probability of desired results while reducing the likelihood of undesirable outcomes, to suggest new questions, identify high-impact interventions, examine new possible actions, and recognize situations in which small interventions might have large effects (and large interventions might have only small effects) (Axelrod and Cohen 1999).

From the perspective of complexity theories, value is placed on bottom-up decentralized approaches, diversity of opinion, risk-taking, learning, adaptation, permanent novelty, and experimentation (Olson and Eoyang 2001). From the perspective of this study, the two most important mechanisms to operate on complex systems are manipulation of variation and selection on the one hand, and the balance between exploration and exploitation on the other.

Managing variety and selection. It was mentioned above that two important mechanisms of CAS are variation and selection; in natural systems, variation is random and the selection criterion is reproductive efficiency. Human interventions operate on both mechanisms, modifying them in a directed fashion, as exemplified by plant breeding (see Box 2.2).

The latter example illustrates a key characteristic of evolutionary processes: contrary to what an engineer would do, the "solutions" to "problems" are obtained through a process of directed search without designing them intentionally. The example of molecular breeding presented in Box 2.3 illustrates the differences between the artificial variation and selection method and the rational design method.[5]

For this mechanism to be useful, variation cannot be infinite. Too much variation raises the cost of gathering and processing informa-

Box 2.2 Manipulating variety and selection

A plant breeder knows the characteristics of the parents available to her and selects those she hopes will pass some desired trait to their progeny (for example, resistance to a given disease). In the early stages of developing a new variety, the breeder usually makes thousands of non-naturally occurring crosses. Thus the breeder increases variety by making crosses she hopes will raise the probability of obtaining the desired result (as opposed to the totally random crosses that occur in nature).

With artificial selection, the breeder overrides the natural process of selection via reproductive efficiency by selecting the progeny that displays the desired properties (for example, disease resistance) while discarding the other crosses without taking into account their reproductive efficiency. Similar mechanisms are being successfully used to develop complex computer programs, synthesize new chemical and medical compounds, and operate on large-scale ecosystems.

[5]These mechanisms receive the name "molecular breeding" because of their similarity to traditional plant breeding (Crutchfield and Schuster 2003).

Box 2.3 Using artificial variation and rational design methods to operate on a CAS

Estrogen is a human hormone that functions by binding onto a specific receptor in cells; estrogen can be thought of as a key and the receptor as a lock. The problem is to find an alternative molecule that can imitate estrogen by binding to the receptor. Today it is possible to easily generate tens of millions of peptides (small proteins each containing a few aminoacids). Each of these peptides is potentially an alternative key that can bind to the receptor.

To find the alternative key, many copies of the receptor are attached to the bottom of a petri dish and are exposed to tens of millions of artificial peptides. The peptides that did not attach to the estrogen receptors are subsequently washed away, and the ones that did bind are recovered. Any of the recovered peptides is a second key that can bind to the receptor and is therefore a candidate to imitate estrogen.

However, biochemical engineers who use a rational design approach start by building a detailed model of estrogen and its receptor, and then design a structure that can serve as an alternative key. The sequence is then synthesized and transformed into a structure with the desired properties. Rational design is clearly a process that is essentially different from the abovementioned process of artificial variation and selection.

The relative efficiency of the methods depends on the complexity and stability of the processes on which one wishes to operate and how much is known about them. If the process is little known, changes rapidly, or is complex, rational design is less effective, because it limits exploration of the possible solution space and bets that the explored solution is the most effective. In these cases, the effectiveness of the rational design approach depends more on luck than the management of variety and selection approach.[6] It has been demonstrated that the method of artificial variation and selection converges on an optimum at least as quickly as the rational design method (Crutchfield and Schuster 2003).

tion while reducing the solution's useful life. However, too little variation can lead to a premature convergence of the system on a suboptimal solution (for example, see Box 2.1).

Interventions aimed at operating on diversity have to balance the creation of diversity at an intermediate point between uniformity and chaotic variation. The fundamental question that decisionmakers face in a CAS is: what should the balance between variation and uniformity be? The response to this question depends on the specific conditions under which the decisionmaker is operating (Kauffman 1995; Axelrod and Cohen 1999).

In a changing world, actors or strategies that are currently suboptimal are valuable resources for the future, whereas uniformity can have grave consequences.[7] For example, farmers only use a few seed varieties well adapted to local conditions (that is, the crop's

[6]An example is the discovery by Monsanto of the gene that confers resistance to the herbicide Gliphosate. Monsanto invested hundreds of millions of dollars over two decades to develop the gene, without success. Eventually, researchers took a sample from a chemical waste sink near a plant producing Gliphosate and discovered bacteria that had adapted to a medium with a high concentration of the herbicide. Soon they isolated the gene that conferred resistance and they introduced it into plants (Charles 2001).

[7]The future value of diversity has been analyzed, for example, in the literature on financial and real options (Dixit and Pindyck 1994) and business administration (Christensen and Raynor 2003; Davila, Epstein, and Shelton 2006).

genetic variation is very limited).[8] With uniformity, farmers obtain the highest yields but create the conditions for the emergence of diseases that are very effective at attacking the varieties in use. Breeders, however, use lower yielding varieties to develop new varieties resistant to emergent diseases.

To manipulate diversity, two complementary strategies may be used: recombination of elements and relaxation of restrictions. Recombination of elements is similar to the work of the abovementioned breeder. This method can be used, for example, to create new strategies, taking one that was successful in a different context and changing certain elements to adapt it to a new situation. In the relaxation of restrictions, the decisionmaker operates on a restriction and observes the response of the system to the intervention. Interventions may be of two types: modifying one of the system's restrictions (for example, developing an irrigation area) or modifying an element (for example, using a new type of packaging for fresh fruits) (Axelrod and Cohen 1999; Rodrik 2006).

Exploration versus exploitation. Policies in complex systems must balance the possibility of choosing among actions or agents that are unproved but might be superior to existing ones (exploration) and actions or agents that are known and for the moment give successful results (exploitation) (Axelrod and Cohen 1999; Davila, Epstein, and Shelton 2006).

Exploitation of proven actions or agents generates the greatest gains but, because eventually these gains diminish, it is necessary to explore and find new sources of benefits. Once again, the balance between exploitation and exploration can be illustrated with plant breeding. Breeders know a group of plants, which they use in their crosses. Thanks to this knowledge, breeders can identify parents with a high probability of conferring given properties on their progeny. Nevertheless, as pathogens evolve, it is necessary to look for new genes that provide resistance to the mutated organisms.

The problem for decisionmakers is to decide the amount of resources that will be allocated to exploitation of the known or to the search for new sources of benefits. Too much exploration reduces the immediate benefits and can generate too many potentially useful products that are lost because of the difficulties in processing increasing volumes of information. Too much exploitation reduces the possibility of finding new sources of benefits when the current ones are exhausted (Davila, Epstein, and Shelton 2006).

Exploration is especially valuable when problems (Axelrod and Cohen 1999)

- are long term or very broad: the greater the probability of obtaining benefits in time or space the more profitable the investment in searching;
- have rapid and reliable feedback mechanisms: if these mechanisms do not exist, it is worth investing in creating them to profit from their exploitation;
- are relatively free of risks of catastrophes stemming from exploration: an example of a high-risk process is research on the smallpox virus, because an accidental escape of the virus could cause a worldwide epidemic; and
- threaten disaster: the benefit from exploration is to avoid the cost that comes from a strategy based only on exploitation. Research on AIDS falls into this category.

A clear understanding of how these instruments may be used is of critical importance for policymakers. For example, the 32 PFs

[8] The genetic variation in this case refers to the variation among seeds and is different from the variation contained in each seed's genetic code (a concept used by geneticists).

serve as a mechanism to explore instruments to foster innovation, but they do not have a system to systematically monitor these experiences; in other words, there is no selection mechanism that allows the PFs to take advantage of variation.

Defining strategies in complex systems.[9] Failures are common in complex systems. For example, it is estimated that in the twentieth century barely 10 percent of companies have been able to sustain growth for long periods (Christensen and Raynor 2003). The failure rate is high because complex processes are difficult to understand and consequently to predict, so that only those companies capable of adapting to changing conditions survive. Few organizations, however, possess this capacity because of the difficulty of modifying structures that have functioned well in the past and that in general are still functioning when the development of alternative structures should be started (Bailey and Ford 2003). Moreover, because the strategies to operate on complex systems are very different from those used in "simple" processes, few organizations posses the knowledge to develop alternative strategies (Christensen and Raynor 2003).

In simple processes, it is possible to obtain a relatively thorough knowledge of the main trends, so the best strategy is to use rational design, that is, to undertake careful planning followed by a careful execution of the plans (see Section 2.1.1). In contrast, in complex processes decisions must be taken with a limited knowledge of the process's dynamics. Therefore, the strategy should not plan implementation but learning. Decisionmakers need to identify what critical information is missing and in what order it will be needed. Planning should reflect these information needs or, at least, should resolve the main uncertainties before committing time and resources to a particular action plan.

Planning in complex processes should recognize that the processes are not well known. With the partial information they possess, decisionmakers construct a mental model of the process. The next steps include (1) questioning the assumptions used to construct the model, (2) using pilot projects to discover the most relevant dynamics and check the adequacy of the original assumptions, and (3) maintaining an adequate reserve of resources to maintain the flexibility needed to correct the course of actions if the initial assumptions are not valid.

Strategic planning and planning by objectives often fail in complex processes, because they do not allow for the discovery of emergent opportunities or for reaction to unforeseen problems. Normally, when the results do not meet the stated objectives, these approaches recommend investing additional resources to close the gap between what was planned and what is obtained. In other words, emphasis is placed on deepening the original lines of action. On the contrary, in complex systems most of the benefits are often derived from unanticipated opportunities. Only when decisionmakers have minds open to identifying these opportunities (that is, they look for them) and enough resources in reserve can they take advantage of unanticipated opportunities.[10]

The PFs operate on especially complex processes. Research is uncertain, because it attempts to develop new knowledge. Moreover the diffusion of innovations is complex, because it depends on many factors, including

---

[9]This section is based mainly on Christensen (2003) and Christensen and Raynor (2003).

[10]Serendipity only happens to those who have a mind open to surprises or who are searching for something, only to find something other than what they were looking for. Alexander Fleming discovered penicillin after bacteria failed to grow in two of several petri dishes in which he was growing them. Instead of washing the dishes and starting the culture again (which probably many scientists did before him), he realized something was preventing the bacteria from growing and investigated the phenomenon.

markets and domestic and foreign economic policies, technological developments, political developments, and cultural elements.

2.2 Innovation Processes

2.2.1 What Are Science, Technology, and Innovation?

To identify what role science should play in innovation processes and to design STI policies adequate for different innovation systems, it is necessary to clarify what science, technology, and innovation are.

In the past decade, a consensus has emerged among philosophers of science that no one characterization of science exists, and no single scientific method has universal application. Each science has its own mechanisms for research and for accepting statements. Furthermore, these statements are based both on empirical observations and on socially acceptable criteria (Machamer 2002). If science is not "a method of discovering objective truths," what is it? This question has different answers, depending on the perspective from which it is analyzed. Following Stoneman (1995), we define science as what scientists do, whereas technology is what technologists do. This definition highlights the fact that the difference between science and technology is not what professionals do but why they do it, and what criteria they use to accept knowledge as valid. In other words, it is recognized that research and development are immersed in social processes that determine what the professionals do.

Scientists seek to create new knowledge, which should be freely diffused as quickly as possible by specialized means. That is to say, the scientists' objective is the creation of information, a public good.[11] In contrast, technologists seek to obtain benefits by creating private goods; thus they seek to keep the information within the firm until it is no longer valuable. The race to decode the human genome exemplifies these differences. Two teams of researchers participated in this race, one coordinated by a private firm and the other by public institutes from several countries. The private team sought to patent the knowledge, whereas public researchers tried to publicize their discoveries as quickly as possible. Both teams were researching the same phenomenon (although they used different methodologies) and obtained the same result, but one was doing science and the other technology.

The other difference between science and technology lies in the acceptance criteria. For science, knowledge is valid when it can explain the phenomenon being studied and the same result is obtained by different researchers working independently. But for a technologist, the criterion is to solve a problem—the comprehension of the underlying processes is a secondary objective. The increasing use of directed search methods to find solutions to "scientific" problems (see Section 2.1.2) and the stronger protection of intellectual property rights sought by public research institutions are blurring the traditional distinction between science and technology.

2.2.2 Innovation Processes: The Importance of Networks in Technology Transfer

An innovation is defined as the successful introduction of a novelty into a social or eco-

[11] A good is defined as public if it is nonexcludable and nonrival. Nonrival means that the consumption of the good by one person does not affect the amount available for other consumers. Nonexcludable means that no one who wishes to consume the good can be prevented from doing so. Open-broadcast television is a public good, because the reception of the signal by one person is not affected by how many television sets receive the same signal (nonrival) and because no person with a television set can be prevented from watching the programs (nonexcludable). Being a public good does not depend on whether it is produced by a public agent or a private firm, but only on these two properties.

nomic process. This definition emphasizes three key elements. First, innovation is the creative use of knowledge in response to social or market needs or opportunities (OECD 1999). It is not enough to invent something; it only becomes an innovation when it is used productively (Fagerberg 2005). An important consequence of the definition is that researchers do not create innovations; they create inventions. Second, the introduction has to be successful: many agents and farmers try new things (Bellon 2001), but not all of these trials lead to practices or products that improve what already exists. Third, innovations are defined as such in a specific social context (Bailey and Ford 2003): what is an innovation in one place or for a specific group of actors might not be in another context (Box 2.4).

In the terminology of complex systems (see Section 2.1.1), innovation results not only from variation (trying new things) but also from selection (finding things that are better than those in use) and its incorporation into economic or social processes. Innovation processes are generally complex (that is, continuous and cumulative) and are at times interrupted by radical jumps. In others words, innovations are mainly incremental and only sometimes radical; but they are always the fruit of technological, administrative, or commercial learning processes combined with a mental attitude open to the identification of opportunities for change (Davila, Epstein, and Shelton 2006).

Innovation processes are complex: if they could be planned in their totality, they would be routines, not innovations (Nickles 2003). Because they are evolutionary processes, their dynamics are determined by the interaction between exploration (including variation and selection) and exploitation on the one hand, and between trends and random processes on the other.

Innovations are classified as product (when they result in something new that is marketed) or process (when they modify the way of doing something without modifying the final product) (OECD 1999). In general, an innovation process includes a succession of product and process innovations (Davila, Epstein, and Shelton 2006). For example, the introduction of a new crop requires the development of new marketing mechanisms, or access to new sales venues requires the adoption of production techniques that produce more uniform crops.

Innovations that have important economic or social impacts only stem from processes that are sustained over time by the combined efforts of actors with different capabilities, because individual actors (including firms) generally do not possess all the resources required to innovate. Thus innovators integrate into innovation networks in which a variety of actors participate (Rycroft and Kash 1999; Christensen and Raynor 2003). Network integration is not easy because of difficulties in implementing collective action (that is, agreeing upon and

Box 2.4 The contextual nature of innovation

In the 1990s, an agronomic package that included planting corn in rows was adopted by many small farmers in Ghana. One component of the package was placing a pole at one end of the plot to help the farmer walk in a straight line (Ekboir, Boa, and Dankyi 2002). Clearly the pole is an innovation only in the context of very poor farmers who do not use machinery. In contrast, improvements in biotechnology techniques used in plant breeding are not innovations for these farmers, because they do not use improved seeds.

implementing rules of interaction, creating confidence, and avoiding opportunistic behavior).

The evolution of innovation networks is determined by the dynamics of the relationships among agents, technologies, markets, and the formal and informal rules that regulate people's behavior. A discussion of the structure and dynamic of innovation networks can be found in Rycroft and Kash (1999), Ekboir (2004), and Powell and Grodal (2005).

The structure and dynamics of innovation networks depend on the complexity and maturity of the innovations and markets. In the case of relatively simple or mature innovations and markets, most innovations are incremental or semi-radical. The networks are more lax; members often relate formally or are mediated by markets, because each agent has a relatively good understanding of the needs of other members of the network and the customers, and of the technological and market opportunities. These are the cases traditionally studied in conventional economic theory and that serve as models for the majority of agricultural programs. Grains, livestock, and traditional export products fall into this category. In contrast, in the case of complex or new innovations, network members have to interact informally, often, and closely to overcome emergent hurdles. Because most of the innovations are radical and/or the markets change rapidly, these networks face great technical and commercial uncertainties, which prevent effective contracting: successful collaborations are based on trust (Rycroft and Kash 1999; Christensen and Raynor 2003). Fresh fruits and vegetables and high-value meat products fall into this category. The characterization of a market or innovation as simple or complex can change, reflecting new technologies or commercial opportunities, as exemplified by gourmet coffees (from simple to complex) or shrimp (from complex to simple).

The effectiveness of innovation networks depends on their capacity to facilitate the exchange of information and resources. In the terminology of Social Network Analysis, this capacity is known as the navigability of the network. An innovation network's navigability depends on the existence of central (well-connected) actors who interact among themselves (Buchanan 2002) and the environment (for example, laws and markets) in which these networks operate. COFUPRO, for example, became a central agent that facilitated the exchange of information among the PFs.

One of the major limitations affecting the innovation capacity of small farmers in developing countries is their inability to integrate into navigable networks that allow them access to technical information, goods, services, markets, and financing. For this reason, one of the central objectives of pro-poor innovation policies should be to strengthen these networks and increase their navigability (see Section 2.3). A few PFs actually played the role of central actors that linked small farmers with processors, technical advisers, and markets (see Section 6.5).

2.3 STI Policies

STI policies have changed over time according to the models used by decisionmakers to characterize these processes. These models can be classified into five generations (Rothwell 1994). The first generation, known as the linear model, comes in two versions. The first version, known as technology push, was developed in the 1940s and holds that results obtained from basic research serve as inputs for applied research, whose output is in turn used for technology development (Bush 1945). According to this model, public STI policies ought to finance basic research in public institutions, stimulate private investment in technological development, and, in certain cases, finance public extension programs. The second version, introduced in the 1960s, postulated that new ideas came from the market, from clients, or consumers (demand pull). In this model, STI policy needed to identify the demands of clients or final consumers and orient research and development activities toward meeting these demands.

This model is still linear because it maintains the assumption that researchers work in research institutions, isolated from other actors in the innovation system. These two approaches do not differ on the role of science in innovation processes but on the origin of the information that researchers should use to define their research programs. In the innovation system perspective, information flows are complex and multidirectional. Technological development often precedes the scientific understanding of the underlying processes (Nelson and Rosenberg 1993; Fagerberg 2005). In the linear vision, researchers pass on knowledge to extension agents, who then pass it to the farmers, who are passive adopters. In the innovation systems framework, farmers receive technical and commercial information from different sources and decide what elements of the proposed package are appropriate to their needs (Ekboir 2002).

The 1970s and 1980s were dominated by the science/technology/market linkage model, characterized by sequential but independent stages and multiple channels of interaction (see, for example, Mowery and Rosenberg 1979). In this generation of models, STI policies should stimulate both technology supply and user demand.

In the 1980s, the discussion focused on an integrated model in which the functional stages (that is, research, development, and adoption) were parallel and highly integrated. In the 1990s, this model evolved into the system/network integration model, in which research was intrinsically linked to technological and economic factors instead of being an autonomous knowledge generator. Thus research activities were nodes of a wide network of knowledge creation and use (Fagerberg 2005). In the final two generations of models, STI policies attempted to balance technology supply with the needs of the market, promoting the creation of networks, fostering change in research institutions and private firms so that they can better integrate into innovation networks, and strengthening innovative capabilities (Lundvall and Borrás 2005).

2.3.1 Innovation as the Engine of Economic Growth

The abundant literature on international "catch up" has shown that success stories have always been associated not only with technical change but also with institutional and organizational change (Fagerberg and Godinho 2005), in other words, with the development of innovative capabilities (see Section 3.1). This finding is confirmed by the fact that within developing countries certain sectors or productive chains have developed unique capabilities (Reardon and Flores 2006), but the growth has not been uniform; in other words, social and economic disparities have increased (Goldberg and Pavcnik 2007). The factors that contributed to the rise of these dynamic sectors have been diverse, but in the vast majority of cases in agriculture, they were not a result of public programs but of the curiosity of individual agents or of private investors (see the case studies in Weatherspoon and Reardon 2003; Gabre-Madhin and Haggblade 2004; and Pardey, Alston, and Piggott 2006). Nevertheless, once the processes were under way, public programs and policies were often influential in their development—but not always in a positive way.

Several technologically backward countries (for example, Denmark, Japan, South Korea, and Switzerland) were able to rapidly catch up with more advanced countries. In the economic growth literature, this phenomenon is known as the "advantage of backwardness" and suggests that those who import or copy a technology can become more competitive than those who developed it (Fagerberg and Godinho 2005). The advantage may stem from the pioneer's problems (for example, the pioneering agent uses older machinery, cannot learn from the experience of others, or faces institutional resistance generated when old technologies are replaced with new, untried ones) or from the follower's advantages (for example, the development of unique capabilities, or exploiting the fact that the technological trajectories have already been identified). In the terminology of complex processes (see Section 2.1.1), the countries that do manage to close the gap initially exploit the international stock of knowledge more efficiently than do other countries and devote fewer resources to exploration. As the country develops, the resources devoted to exploration should increase (Fagerberg and Godinho 2005).

The experiences of those countries that have not developed clearly show that importing modern technology or investing in primary and secondary education does not guarantee success (Fagerberg and Godinho 2005). Successful development requires an environment that favors innovation and agents that have the capacity to absorb and digest knowledge generated elsewhere. Countries that succeeded in catching up dedicated substantially more resources to the acquisition, assimilation, and adaptation of imported technologies than to the direct purchase of technology (in some cases, the investment ratio was 3:1) (IDRC 1997). In addition, the Asian Tigers combined significant capital and infrastructure investment with reverse engineering and experimentation in selected industrial sectors and an increasing supply of highly qualified workers and professionals (Fagerberg and Godinho 2005).[12] In contrast, the Southern Common Market (MERCOSUR) countries also made important investments in public research institutions and imported technology in the 1960s and 1970s but did not make the effort to assimilate and learn from these importations.

If companies do not invest in developing their absorptive capabilities, it becomes impossible for them to use the knowledge generated by the research system and by other companies (Davila, Epstein, and Shelton 2006; Dosi, Llerena, and Sylos Labini 2006). Neither can they use trained professionals to assist them, because they cannot identify their technological and innovation needs.

2.3.2 Relationship between Research and Innovation Systems

Innovation systems comprise networks of actors (including farmers, firms, universities, research centers, technological institutes, training centers, and financial institutions), their actions and interactions, and the formal and informal rules that form the institutional framework in which the networks operate. The actions and interactions influence the production, diffusion, and use of economically and socially useful knowledge; in other words, they affect the innovative performance of companies and social processes (Friedman 1987; Lundvall 1992; Nelson 1993). Studies of innovation systems are focused on the analysis of the interactions, the flows of information and knowledge between system agents, and the actions to intensify these flows and interactions (see Section 2.2).

The innovation system is larger than the research system, because only research institutions participate in the latter. The analysis of the larger system is necessary because:

1. Scientific research only has value when it improves productive systems and life qual-

[12]The Asian Tigers is a group of fast-growing countries that includes Hong Kong, Singapore, South Korea, and Taiwan. MERCOSUR is a customs union formed by Argentina, Brazil, Paraguay, and Uruguay as full members and Bolivia and Chile as associated members.

ity; that is, when it is used to satisfy social or economic needs or take advantage of emerging opportunities.

2. Technologies are becoming more complex not only because they are made of more components and their development requires a greater scientific effort, but also because agents must compete in increasingly integrated markets that are changing at an accelerating pace (Rycroft and Kash 1999; Dosi, Llerena, and Sylos Labini 2006). Products that were previously produced with simple and/or well-known technologies and sold in stable local markets are being progressively integrated into the global economy (Berdegué 2005).
3. The adoption of innovations is a creative and interactive process that involves both market and nonmarket institutions. Furthermore, innovations are increasingly dependent on effective interactions between a country's scientific base and its business community (OECD 1999). As the scientific base of technological development has grown, the information that permits development of new technologies is ever more in the public domain but also increasingly difficult to understand (Nelson 1998). Investment in human capital (especially graduates from foreign universities) is an efficient tool for building the capacity to take advantage of knowledge developed in other countries.
4. Companies are today the main source of innovations. Their success depends on the incentives found on the economic and regulatory environment, access to critical inputs, the internal capacity to take advantage of economic and technological opportunities, and their abilities to integrate themselves into innovation networks (Christensen and Raynor 2003).
5. From the perspective of the innovative agents, scientific activities are more important as a source of highly qualified personnel and scientific information than as generators of commercially relevant information (Mowery and Sampat 2005; Dosi, Llerena, and Sylos Labini 2006). When researchers interact actively with innovating agents, the former help the latter to strengthen their absorptive capabilities.

2.3.3 Balance between the Creation of Information (Research) and the Use of Existing Information (Innovation)

The balance between the use of existing information and the creation of information new to the world is one of the most important issues for the design of STI polices: the problem is that there are no objective procedures to determine this balance. The innovation system being complex, the balance depends on a set of factors, especially the sophistication of the innovation system, the patterns of interaction among different agents, and the political economy of financing research and development activities (Branscom and Florida 1999; Kraemer 2006).

In developing countries, one of the factors that most influences this balance is the negotiation of public administration budgets, because it defines the investments of public research institutions in physical and human capital. However, these negotiations rarely include an explicit discussion of the countries' scientific and technological needs; in other words, policies in these areas are defined implicitly by a small group of agents with a limited understanding of the subject.

The sophistication of the innovation system defines the limits and possibilities of innovation policies. Within these policies, one of the most important decisions is the balance between domestic development (exploration) and imitation or adaptation of information generated in other countries (exploitation).

Countries or sectors with weak innovation systems depend almost exclusively on technologies developed in other countries. In these cases, investments in formal research and development have little influence on growth rates, and investments in education and training seem to play a much more significant role (Gittleman and Wolff 1998). The reason is that education and training programs increase the capacity to explore the largely unknown international stock of information. But these policies are only effective if the society can productively use the trained manpower, for which it is necessary to establish policies to strengthen the absorptive capabilities of private companies and the public sector. Otherwise the most capable people migrate in search of better working conditions.

Weak research institutions can only undertake simple research (that is, minor developments). To tackle more complex tasks, it is necessary to strengthen these systems through investments that are significant and sustained for prolonged periods, until the critical mass necessary for effective research institutions is reached. In addition, it is necessary to create institutional cultures that enable the expression of the researcher's creativity. Thus investments aimed at creating original information are only beneficial in the long run, and only if the information can be transformed into knowledge and innovation.

Beyond a certain level of development, the strategy of importing technology without a local adaptation effort becomes ineffective (Malerba 1993; Archibugi and Michie 1998). At that stage, it becomes necessary to carry out local research and development activities; these activities have special characteristics (Box 2.5) which make them complex. Therefore, specially designed public policies are needed.

To increase an economy's competitiveness, it is necessary to strengthen innovation networks through long-term policies to (1) build up research institutions; (2) foster research and development in private companies; and (3) promote investment in human resources, research infrastructure, and the creation of instruments that promote interinstitutional collaboration on research and development (Lundvall and Borrás 2005). The success of these policies depends on continuity, because institutions that required many years to build can be destroyed rapidly if adequate incentives are not maintained.

As the innovation system gains strength, the research spectrum widens. Eventually, it can reach a high level of sophistication. Once this level is reached, the country or sector no longer depends on imported technology but can now develop its own. The stronger the research system, the more important the international search for knowledge becomes, so that promising lines of research can be identified and competitiveness maintained (see Sections 2.1.2, 2.2.1, and 3.1). In other words, as the innovation system evolves, the profile of research institutions, the private sector's attitude toward knowledge absorption, the competence of technicians, the private-sector links with research institutions, and the government's capacity to design appropriate STI policies must also change in response to the new conditions for knowledge generation (Fagerberg and Godinho 2005).

In developing countries, universities are generally weak and work in isolation from other actors in the innovation system (Eun, Lee, and Wu 2006). In addition, they have an organization similar to most European universities, where these institutions are supposed to research, educate, and link with the private sector. It has been found, though, that this arrangement has not resulted in first-class institutions in any of the three tasks they are supposed to address (Dosi, Llerena, and Sylos Labini 2006).

Technological institutes can be an essential instrument in support of the private sector in the identification and digestion of useful information. To do this, they may play a more important role than universities, because they have been created to undertake applied research and to liaise with the private sector (although they have not always done so), and because the linear vision that prioritizes basic research (and is isolated from the rest of soci-

Box 2.5 Features of research and development activities

1. Research and development activities are more uncertain than most other activities in terms of the probability of obtaining results, their nature, and the timeframes in which they can be productively used. Often the results obtained were not foreseen at the beginning of the research process and turn out to be more valuable than was originally aimed for. For example, the Post-it resulted from using a nonsticking adhesive (which was considered a failure in the original project) to design a new product (Davila, Epstein, and Shelton 2006). Furthermore the total value of a new discovery may be unknown even at the end of a project (Huffman and Just 2000; Christensen and Raynor 2003).

2. More than one approach may be valid to study a given problem, and it is impossible to predict which is the most appropriate. Similarly there are alternative solutions to the same problem, and not all of them come from formal research (Huffman and Just 2000; Kraemer 2006). For example, losses in the transport of fresh fruit can be reduced by improving roads, developing new packaging methods, or creating a variety more resistant to shocks.

3. Most of the benefits from investing in research and development come from the best discovery and not from the sum of all the discoveries. For example, if various crop varieties adapted to a particular ecosystem are developed, only the best variety will be adopted by farmers. Thus, much of what is discovered does not have immediate social value, because it is not substantially better than the best available technology (Huffman and Just 2000).

4. Even research projects that do not obtain the desired results or are not used in productive processes can be valuable, because they provide information useful for defining other research strategies, or they can prove valuable in the future. The "failed" projects provide information about alternative research strategies and thereby increase the probability of success of new projects (Weisberg 2003). In addition, results that have no applications today can be valuable in a different social and economic context.

5. The relationship between a scientist and the administration of the institution that employs him or her is asymmetric, because it is costly (and probably ineffective) to control the scientist's effort (Huffman and Just 2000). Given the uncertainty at the beginning of a research project, the real effort put in by the scientist cannot be inferred from the results. Researchers that carry out the most innovative or most risky studies have a lower probability of obtaining results than do researchers conducting more conventional work. But the former may produce better quality work than the latter, even when the results are not what was aimed for. An incentive structure based on outputs discriminates against the most innovative or risk-taking research projects—the very ones who often produce greater mid- and long-term benefits (Davila, Epstein, and Shelton 2006).

6. Scientific productivity, measured by publications, has an extremely asymmetric distribution. For example, bibliometric studies show that the great majority of published works are never cited, but this does not imply that they are of no importance, because they can influence other researchers in ways that are not registered by these studies. Furthermore, one study found that in North American universities, 50 percent of published work is produced by 10 percent of researchers (and fully half of that work is published by only 2 percent of the total number of researchers). The other 50 percent of published work is produced by the remaining 90 percent of researchers (McClellan and Dorn 1999). Research incentives based on this indicator would exclude most researchers. There are also consequences for interinstitution comparisons. If researchers could move freely, the institutions that publish the most would be able to invest more resources in hiring the "most productive" researchers, creating two categories of institutions. If there is no free mobility and the researchers are tenured from the moment of being hired, publishing capacity is a random variable that is resolved at the moment in which the institution contracts a researcher over whom it will subsequently have very little control.

ety) has generally predominated in universities. The principal role of the universities ought to be preparing highly trained professionals (Mowery and Sampat 2005).

2.3.4 Extension Systems from the Innovation System Perspective

From its beginning at the end of the nineteenth century, the mandate of public extension was to pass information from research institutions to farmers. Over the years, this mandate expanded, and today extension agents are expected to have a broader vision of agriculture and are no longer considered simply as channels for the passage of information but also as consultants, facilitators for collective action, and conduits of information flows (Alex, Zijp, and Byerlee 2002).

Following these changes in the vision of extension, various countries have experimented with a variety of institutional arrangements for public extension: decentralization (within the same organization), geographical dispersal, outsourcing, public–private alliances, and privatization. In parallel, nongovernmental organizations (NGOs), farmer organizations, and private firms have also begun to offer extension services. In most cases, the institutional arrangements that have involved public organizations (even when they are only funding private extension) have continued to reflect the linear vision of science, in which the extension agent essentially fulfills the role of information channel (Rivera and Alex 2005). In contrast, in both developed and some developing countries highly effective private-extension mechanisms have emerged.

From the point of view of the innovation systems framework, extension agents need to play a different role than that assigned in traditional models. As is explained in Section 3.1, innovation depends on the individual and collective capabilities of agents to absorb information and to use it in the development of solutions to their economic or social problems. In this conceptual framework, the extension agents should function as facilitators throughout the innovation process. In other words, the extension agents must not focus simply on "passing a package," but instead their interventions should adapt as the innovation process matures.

Most extension agents have technical training related to agriculture. Nevertheless, if it is accepted that the biggest difficulty that farmers (especially small ones) have is the accumulation of human and social capital and integration into innovation networks, then it is not necessary that extension agents know about agriculture, but rather about how to facilitate social processes and collective action. Thus extension agents should be social workers or educators rather than agronomists or animal-production specialists. Traditional extension agents specialized in technical subjects can only help those farmers who are able to absorb external knowledge and use the extension agents' advice; generally these are the largest or the most innovative farmers.

Recognition of the new roles extension agents should play and what these roles imply for the organization of extension programs and the training of professionals is of key importance for the functioning of the PFs and public extension programs.

2.3.5 STI Policies

STI policy recommendations changed as new models of the innovation processes gained recognition. Nowadays it is accepted that these policies must adjust to the level of economic development and the strength of research and extension systems (see Sections 2.3.1 and 2.3.2).

In developed countries with consolidated innovation systems, the prime objectives of scientific policies are to (1) expand or reach the frontiers of scientific knowledge, (2) train high-level human resources, and (3) contribute to the solution of social and economic problems (EC 2003). To attain these objectives, it is necessary to consolidate research capabilities, adapt the intellectual property regime, and encourage the adoption of criteria of excellence (Lundvall and Borrás 2005; Dosi, Llerena, and Sylos Labini 2006).

In turn, the goals of technology and innovation policies are to (1) generate technological knowledge, (2) promote the transference and assimilation of technology in strategic areas, (3) foment the learning capacities of companies, and (4) train human resources. Finally, innovation policies have as objectives to (1) raise the rate of successful introduction of new products, processes, and services; (2) support the creation and diffusion of knowledge, especially in support of private research and development; and (3) contribute to economic growth and increased productivity and competitiveness (Lundvall and Borrás 2005).

Developing countries, however, generally have relatively weak research systems (World Bank 2003; Eun, Lee, and Wu 2006). For this reason the emphasis should be placed on strengthening the innovation system while building the research system (see Section 2.3.2). Policy instruments must vary with the capacity of the innovative agents and research institutions. Furthermore, it is likely that the capacity of all these agents will not evolve in parallel, so specific sets of policy instruments should be developed for specific situations.

Traditional scientific research stresses the creation of information that is new to the world. In contrast, innovations mostly use existing information to improve a social or economic process (Fagerberg 2005). In the process of absorbing information, sometimes new information is created. In other words, innovation processes can also generate scientific information. The main problem for countries and sectors that are not on the technological frontier is to balance the creation of new information and the use of existing information. This balance determines what instruments to use in STI policies.

According to the linear vision of science (see above), the public sector must support research to compensate for the private sector's suboptimal investment (from a social point of view). The underinvestment is the optimal response to

- indivisibilities that arise because there is a minimum level of investment below which research is not efficient;
- inability of the private sector to appropriate all research benefits because scientific information is a public good; and
- uncertainty and the long time required for investment to mature, which affect the private sector more than the public.

The innovation systems framework shows that these justifications are insufficient, because they are based on false assumptions about the behavior of firms on the one hand and the relationship between the creation and use of technical information on the other. Recent research on innovative capabilities has shown that the public-goods justification is based on a simplistic view of research and technology. In fact, all public information is a public good, but that does not mean that private agents can use it. Pure public goods do not require any special effort or skill on the part of the recipient of the services or goods. But this is not the case with technical information: to use it, innovative agents have to invest substantial resources to develop absorptive capabilities that enable them to identify and digest existing information. In other words, although public goods are free, their use is not, and spillovers only exist when agents invest in their absorptive capabilities (Fagerberg 2005; Verspagen 2005). The difference in absorptive capabilities (see Section 3.1) explains why some agents succeed where others fail.

In addition to the traditional justifications, the state should intervene to solve systemic problems. In particular, the state should not only define the quantity of resources dedicated to science and technology policies but also help improve the functioning of the innovation system. Another fundamental difference between the traditional vision of science and that of the innovation system is that the former emphasizes technological change in isolated firms, whereas the latter emphasizes that innovation results from interactions among agents (Lundvall and Borrás 2005).

Besides market failures, the difficulties that firms have in innovating come from insufficient infrastructure, problems of transition to new research routines, inertia stemming from technological trajectories, cultures that do not favor interaction with other agents in the innovation system, institutional failures, government failures, and systemic failures. The latter include problems in the capital markets, when there is very little and very expensive venture capital, and in those markets for highly qualified human resources that lack a critical mass of professionals for STI development. Government failures are associated with deficiencies in the allocation of resources among different instruments and agents, and with distortions in the incentives offered to researchers that discourage behavioral changes aimed at excellence, pertinence, innovation, linkage, and regionalization. The systemic failures are linked to the lack of an interactive and comprehensive understanding of the functioning of the system (Foro Consultivo 2006).

In the same way that globalization reduced the ability of countries to implement totally divergent economic policies, the integration of innovation systems is intensifying the pressure to harmonize national intellectual property rights and economic, political, tax, and scientific policies. This trend toward convergence does not mean that governments have lost their capacity to influence economic development, but rather that the dynamics of the international and national economies have changed along with the available policy options (Fagerberg and Godinho 2005). In the specific case of scientific policies, it is now recognized that active participation in the world economy increases the potential for economic development, as it (1) raises the technology supply available to domestic companies, (2) improves access to the stock of international knowledge, and (3) elevates the standards for scientific and technological activities (Nelson 1993; Archibugi, Howells, and Michie 1999; Fagerberg and Godinho 2005).

Analysis of the international experience helps identify key elements that countries not on the scientific frontier should consider in the design of scientific and technological policies:

1. One of the fundamental conditions for strengthening an innovation system is defining the role that each institution should play (Ekboir 2004; Dosi, Llerena, and Sylos Labini 2006), in other words, defining which institutions will become centers of excellence in research and which will become teaching centers. Once this decision is made, institutional goals, the resources necessary to attain them, and the incentives offered to professionals should be changed accordingly. In Mexico, research centers include both Public Research Centers

and those in the private sector (for profit and nonprofit).[13] Teaching institutions related to the rural sector include public and private universities and technical colleges. Most educational institutions have research centers, but the vast majority of these have very limited research capabilities (Ekboir 2004).

2. Research institutions should play an increasingly important role in the consolidation of the innovation system (Fagerberg and Godinho 2005). For them to play this role, it is necessary to define strategies to raise them to international scientific standards.[14] Changes in institutional cultures and interchanges of professionals with other domestic and foreign institutions should figure prominently in these strategies.
3. The complexities of public policies, including STI policies, are growing. Governments increasingly have to rely on scientific information to regulate the acquisition and absorption of technology and to improve their own activities. For example, regulating the use of genetically modified organisms or analyzing the environmental impact of agricultural activities requires sophisticated analytical abilities that are normally possessed only by researchers. However, to make use of this information, public officials must also develop their own capabilities.
4. The generation of scientific information by research institutions is not sufficient for it to be adopted as a technology. In the case of the rural sector, extension programs, the farmers' innovative capabilities, and the regulations that link research institutes with farmers greatly influence the dynamics of the innovation system. To foster innovation, extension programs must adopt new modes of interaction with researchers and innovative agents (see Section 2.3.4).

In this context, scientific and technology policies should seek to strengthen the capacity of the system to use and generate information useful for innovation agents, especially (1) improving researchers' capabilities, (2) integrating them into international research networks, (3) transforming institutional cultures and incentives to facilitate the integration into innovation networks, (4) promoting the entrance of young researchers into the system, and (5) consolidating the research infrastructure.

The most important innovation policies are those that seek to increase (1) interactions among the system's agents and institutions; (2) information flows and identification of useful foreign and domestic innovations; (3) the private sector's innovative capabilities, including financial support for its innovative activities; and (4) domestic supply of inputs key to the processes of research and development, especially research infrastructure and the training of scientists and highly specialized workers (Lundvall and Borrás 2005).

Innovation policies cover an ample range of topics and include educational and research financing policies, financial regulations, tax law, market competition and intellectual property laws, support programs for technological prospecting by private firms, and strengthening the public sector's capabilities (Arnold et al. 2000). These policies can be divided into several broad categories (Table 2.1).

2.3.6 Mechanisms to Define Research Priorities

Among researchers of STI policies, there are basically two approaches to define research priorities in the public sector. In the traditional approach, priorities are defined by the organizations that manage the funds, and then

[13]The nature of Public Research Centers as legal entities is defined by federal law. The law allows them greater independence in managing their resources than traditional public offices do.

[14]Research that does not have high standards of quality is not research. But besides maintaining high quality, researchers must interact actively with other agents of the innovation system.

Table 2.1 Typologies of innovation policies

Category	Examples of policy
Foster the emergence of an innovation-friendly environment	Support the formation of human resources in science, technology, and innovation Reduce key market failures, such as financing of research Maintain macroeconomic and social stability
Foster the emergence of a culture of innovation	Diffuse best practices in innovation management Update the curricula of public educational institutions Promote a change of mentality and culture in public research institutions
Foster the spread of innovations	Help strengthen private firms' absorptive capabilities, for example, with consulting services, extension activities, technological interchanges, and seminars and workshops in marketing and administration Improve interactions between public research institutes and firms, especially small and mid-sized ones Use the state's purchasing power to create markets for new products Foster the emergence of bridging institutions to facilitate interaction between productive and social agents and researchers
Encourage the emergence of innovation networks	Act as catalyst, bringing together potential partners Diffuse information about commercial and technological opportunities Support researcher mobility between public research institutes and private companies
Strengthen research and development capabilities	Increase investments in research Promote adjustments in public research institutions, especially in incentives and quality standards Promote public–private alliances for research and development Promote international scientific and technological interchanges Promote labor mobility of scientists and technologists
Respond to globalization	Promote networks of nationally competitive firms Promote international alliances among companies Strengthen the research system, ensuring it is permanently updated and is inserted into international knowledge networks
Improve the design and implementation of policies	Strengthen the analytic capacity in the public sector Improve communication between the public sector and the rest of society Ensure that users of policies participate in policymaking process

Source: Elaboration based on Arnold et al. (2000).

a system of competitive funds is used to distribute the available resources according to agreed priorities (Alston, Norton, and Pardey 1995; Echeverría and Elliot 2002). In spite of the stated objective of guiding the actions of public research institutions, the experience of developing countries shows that these mechanisms have not functioned as expected (Gill and Carney 1999; Vera-Cruz et al. 2008).

It is worth pointing out that most developed countries and advanced research centers—as well as companies with strong research programs—lack centralized and rigid mechanisms for priority setting. In a survey of 95 research universities in the United States, Welker and Cox (2006) found that 65 percent of researchers thought that their institutions established overall institutional priorities, but only 35 percent thought that their institutions defined research priorities. Moreover, they agreed only weakly that research priorities were known by faculty.

The strength of these research systems is not based on a clear definition of priorities but rather on the quality of the researchers, a diversified research system (that is, a system with multiple capabilities that can be tapped for specific research needs), and strong interactions with the users of the technologies (Nelson and Rosenberg 1993; Kraemer 2006).

The traditional approach for defining research priorities has been criticized because the rigid definition of priorities at a national level may leave out research opportunities that do not have a national scope or are not sufficiently well understood by those who manage the prioritization process (because the processes they are expected to influence are very complex or the main trends have not sufficiently consolidated) (Teubal 1997; Huffman and Just 1999; Christensen and Raynor 2003). In the terminology of complex systems (see Section 2.1.2), the traditional approach would be classified as rational design. As seen below, the alternative approach is based on the management of variety and selection.

In recent decades, several developing countries have established research-prioritizing mechanisms through the identification of user demands (Byerlee, Alex, and Echeverría 2002). The experience of the past two decades has shown that these mechanisms have not solved the problems affecting research systems. The reason is that demand-driven mechanisms continue to be structured according to the linear vision of science and do not solve the problem of interaction between researchers and users and, additionally, do not influence research quality.[15]

The second approach, based on complexity theories (see Section 2.1.2), is based on the premise that the definition of research priorities (a complex process) requires more flexible mechanisms than just demand-identification. These mechanisms should recognize the complexity of the systems they seek to influence and the capabilities of all actors in the innovation system. The process of prioritizing research, then, is organized as an exploration to balance the exploitation of existing knowledge and capabilities and the exploration of new opportunities (Branscom 1999; Christensen and Raynor 2003; Davila, Epstein, and Shelton 2006). A successful example of opportunity exploitation is the Fundacion Chile, a nonprofit privately owned corporation created in 1976 by the Chilean government and the ITT Corporation of the United States. The Fundacion successfully supported the joint development or adaptation of technological and business models for several products that captured significant portions of international markets, such as salmon and fresh fruits.

2.3.7 The Effectiveness of Competitive Funds

The mechanisms to disburse public funds and the amounts invested determine to a great degree the dynamics of research systems. In most developing countries, the amounts dedicated to public research are decided as part of the discussion of the national budget, without consideration of the relevance of the research carried out with these resources.[16]

The deregulation of national economies since the 1980s, globalization and the consequent pressure on competitiveness, the reduced effectiveness of traditional policies in the new international context, and the growth experience of various developing countries have forced a re-evaluation of the administration of science and technology. In developed coun-

[15]Hall et al. (2001) relate the experience of a cooperative of mango farmers in India. Needing to solve technical problems, they contracted the services of three universities. The researchers developed the technological solutions, but during that process they did not interact with the farmers. The result was that the "solutions" were not adapted to the resources that the farmers had and so were never used.

[16]In a few developing countries (for example, Brazil) the states also contribute to the financing of research.

tries, direct government financing has fallen and new mechanisms have been introduced to finance innovative activities, including the creation of competitively awarded research funding, contracts for the development of specific products, purchases of new products by the public sector, subsidies for innovation activities in private firms, and the formation of public–private consortia (Echeverría 1998; Branscom and Florida 1999; Huffman and Just 1999; OECD 1999; Lundvall and Borrás 2005; Kraemer 2006). Even competitive funds vary in their structure and selection mechanisms (see, for example, Lepori et al. 2007; Thèves, Benedetto, and Larédo 2007).

These new mechanisms did not replace traditional financing but rather complemented it. In the United States, the share of competitive mechanisms in total financing of agricultural research is limited. In 2000, they accounted for less than 4 percent of the financing of state experimental stations, whereas direct federal and state budget contributions accounted for 74 percent of the total; private contributions, 15.3 percent; and other sources, 6.7 percent (Huffman and Evenson 2003).

Although there is a consensus among specialists that the financing of public research institutions should combine fixed budget contributions with variable allocations (Echeverría 1998; Huffman and Just 2000), in the developing world competitively awarded financing has been almost the only instrument used to allocate operational funds. The arguments in favor of these mechanisms (Echeverría and Elliot 2002) are that competitive funds

1. increase research effectiveness by transferring funds to the more productive researchers;
2. increase efficiency by cost reduction through competition and cofinancing schemes that reduce duplication of effort, exert more control over research activities, and allow better utilization of installations by providing operational funds;
3. promote the identification of research priorities at the national level;
4. increase the flexibility of fund allocation by prioritizing new research areas or pending problems;
5. promote a demand-driven and object-oriented research system;
6. boost interinstitutional interaction;
7. encourage diversification of the national innovation system by financing scientists from institutions that traditionally received no public funding;
8. mobilize additional funds;
9. bring about institutional change in the national innovation system by separating scientific financing from operational policy; and
10. augment research quality, because researchers would know their reviewers' comments.

Among the identified problems of competitive funds are (Axelrod and Cohen 1999; Huffman and Just 2000; Echeverría and Elliot 2002; Ekboir 2004; García and Sanz-Menéndez 2005; Laudel 2005; Kraemer 2006; Vera-Cruz et al. 2008)

1. long-term research capabilities may be compromised, because in general only operational costs are funded, preventing required investments in infrastructure and equipment;
2. allocation mechanisms in small research systems cannot be transparent, because the researchers can identify the reviewers by their comments and writing style—thus reviewers may be reluctant to criticize a proposal, knowing that in the future they may be evaluated by those whose proposal they are currently evaluating;
3. academic quality may not be the most important factor in the decision to fund the project;
4. funds finance relatively short-term projects, biasing resource allocation against

multiyear projects and so discouraging fundamental research;

5. they do not finance either institutional strengthening or investment in human capital;
6. they increase uncertainty about financing;
7. they are biased against novel approaches, because reviewers do not want to be blamed for approving untested methods that can fail;
8. they have high transaction costs, because much time is invested in the preparation of proposals, and in some cases, in lobbying the funding agency, thereby reducing the effective research time;
9. they can be biased in favor of large institutions;
10. research problems concerning a single state or region may be underfunded in the national competitive-grant process, even though such problems could be critical for those areas and may have a large net social payoff;
11. uncertainty is increased when funds are awarded to lesser known institutions; and
12. except for a few large research systems, proposals from different disciplines usually compete among themselves, reducing the transparency of the selection process, because there are no unequivocal methods to choose among two good proposals from different disciplines.

Although there are indications that competitive funds have had major influences on the way agricultural research is conducted, solid empirical evidence of what these impacts have been is almost nonexistent (see below). Moreover, most of the few publications that analyze competitive funds take for granted that they are an efficient instrument to allocate research funds. This point is crucial, because it has been long recognized that the effectiveness of competitive funds depends on the presence of appropriate research capabilities (Gill and Carney 1999). Strong anecdotal evidence, however, indicates that not even developing countries with large research systems (for example, Brazil, India, or Mexico) have reached a threshold where competitive funds can operate effectively.

Based on the assumed effectiveness of competitive funds, several authors have analyzed their impact on the research system, including redirection of research activities, changes in governance, transformations in research institutions, institutional and individual reputation building, increased interinstitutional collaborations, and impact on research competitiveness (Echeverría and Elliot 2002; Toro and Espinosa 2003; García and Sanz-Menéndez 2005; World Bank 2006b), without realizing that most of these effects can also be obtained with other allocation mechanisms (for example, non-bid contracts).

Traditionally, evaluation of proposals submitted to competitive programs has been based on academic quality as measured by peer review. Interviewing researchers in Germany and Australia, Laudel (2005) found out that the quality of proposals was an important but not sufficient factor in the allocation of competitive funds. Other factors, such as the policies implemented by the research institutions and previous success in obtaining funding, influenced the probability of obtaining a grant. This finding is supported by Rubenstein et al. (2000), who found that competitive funds were the most concentrated mechanism of all those used to fund agricultural research in the United States.

Competitive funding is a preferred mechanism in Mexico for the allocation of public resources to cover the operating costs of research and innovation projects. The PFs have used them for several years, but their implementation has evolved as the foundations learned from experience. Moreover, frustrated because they felt competitive grants were not effective in fostering innovation, the PFs started to try alternative approaches to achieve this goal.

CHAPTER 3

Innovative Capabilities, Organizational Learning, and Poverty Alleviation

Throughout their existence, the PFs accumulated experiences and developed capabilities that enabled them to become key actors in the Mexican agricultural sector. This evolution was characterized by the interaction of a few highly innovative individuals operating in an environment that enabled expression of their creativity. Section 3.1 analyzes the nature of innovative capabilities, the interaction between individual and collective innovative capabilities, and how these capabilities influence institutional learning. In addition to capabilities, an organization's innovative performance depends to a great extent on its institutional culture and the structure of governance it develops. Some essential features of organizational cultures are discussed in Section 3.2, and governance is analyzed in Section 3.3. Finally, Section 3.4 discusses poverty-alleviation policies from the perspective of innovation systems, especially how these policies should be adapted to reflect a better understanding of innovation capabilities.

3.1 Organizational Innovation Capabilities: The Role of Learning

Research, development, and innovation enable organizations to improve their capacity to absorb information generated by other agents and to use new internally created knowledge, technological as well as other kinds (Tidd, Bessant, and Pavitt 1997).

Innovative organizations possess a number of characteristics that can be grouped into strategic and organizational capabilities. The former include long-term vision and the ability to identify, process, and assimilate economic and technological information. The latter include internal and external cooperation with Public Research Centers, clients, and suppliers, and the ability to become involved in processes of change and learning. Human resources are paramount in the innovation process (Dosi, Nelson, and Winter 2000).

Just as individuals may or may not have the profile of an innovator, organizations may or may not develop innovative capabilities. Organizational innovative capabilities are a function of individual factors (such as individual creativity) and collective factors (such as work routines, collective learning mechanisms, and organizational cultures) (Christensen and Raynor 2003; Davila, Epstein, and Shelton 2006). For example, the Xerox laboratory in Palo Alto, California, developed in 1973 a personal computer with a graphic interface and mouse (subsequently copied by Apple), an operating system that enabled the simultaneous use of multiple applications (subsequently sold to Bill Gates and used as the basis for Windows; see Box 2.1), Ethernet connectivity, and the first word processor that showed directly on screen the document as it would be printed. But the company's culture, focused on photocopiers and big computers, did not allow Xerox to profit from these developments (Carayannis, Gonzales, and Wetter 2003).

The traditional vision of the organization is based on the assumption that it optimizes its operations subject to exogenous restrictions. The resource-based vision of the organization, in contrast, posits that it obtains benefits by combining its idiosyncratic resources to create new capabilities that effectively relax external restrictions (Henderson and Cockburn 2000; Christensen and Raynor 2003). Capabilities are defined as the ability to carry out planned actions. To be capable of something is to have a generally adequate and reliable capacity to accomplish the objective as a result of a deliberate action. Capabilities bridge the gap between intention and results, and they do it in such a way that the results are definitely related to what was attempted (Dosi, Nelson, and Winter 2000).

Capabilities imply organized activity. In the specialized literature it is recognized that organizational capabilities reside in individuals (including managers, technical support personnel, and line workers), the technology, and the structure of the organization (including routines, methods of coordination, and learning processes) (Argote and Darr 2000).[1] Among these factors, the organizational principles through which individual and collective skills are structured, coordinated, and communicated play an important role. That is, organizations are social communities that use their relational structure and shared codification methods to define transference and communication mechanisms for new capabilities (Zander and Kogut 1995) and to condition individual innovative abilities (Bailey and Ford 2003).

An alternative view of the nature of organizational capabilities posits that these result from the interaction of resources (individuals and fixed capital), processes (the mechanisms agreed on by the organization for doing things), and values (including institutional culture and long-term goals). In new organizations, most capabilities reside in the resources, especially their people. The incorporation or departure of a key person can have a big influence on the probability of success. In time, the capabilities of successful organizations are transferred to processes and values (Christensen and Raynor 2003).

It is useful to distinguish between technological capabilities and organizational capabilities, although the two concepts overlap in the real world. The distinction is based on the fact that technological capabilities refer to elements of scientific and technological knowledge, and to routines essentially concerned with the structure of nature and how to manipulate it (for example, transforming a piece of metal into a particular structure). Organizational capabilities, in contrast, are shared elements of knowledge and routines concerning governance, coordination, and social interaction in the organization and with external entities (for example, with suppliers and clients) (Dosi, Coriat, and Pavitt 2000).

In addition to being an essential part of the organizational memory and production processes, routines reflect governance structures, control methods, decentralized decision-making mechanisms, and compromises among divergent interests. In essence, routines are conscious or unconscious agreements that arise from conflicts of interests in the organization (Coriat 2000). At times routines are so com-

[1]Routines are an essential component of organizational capabilities. The development of new capabilities is based, to a great extent, on the adoption of routines more effective than the currently used ones.

plex and tacit that the organization itself is not aware of their existence or does not understand them (Dosi, Nelson, and Winter 2000).

Capabilities are specific to each organization—their development is based on investment and idiosyncratic processes, and other organizations have difficulties in imitating or understanding them (Teece, Pisano, and Shuen 1997). That is why some organizations succeed where others fail. Capabilities are also contextual: those that confer advantages in a particular context can become handicaps in a different situation (Appelyard, Hatch, and Mowery 2000; Christensen 2003).

Because organizational capabilities cannot be easily bought, they must be built through sustained investments, strong leadership, and adequate selection of those responsible for new projects (Christensen and Raynor 2003; Davila, Epstein, and Shelton 2006). However, organizations cannot develop their capabilities arbitrarily; instead they follow specific trajectories that define the options currently open to the organization and limit future options. In the long run, decisions to develop specific capabilities are almost irreversible (Teece, Pisano, and Shuen 1997; Dutrénit 2001).

An essential issue faced by all organizations is how to maintain, nurture, and renovate existing capabilities so that they can respond to changes in the business and technological environment (Leonard-Barton 1995). Efficient development of new capabilities requires a vision of the changes to be introduced, many trials to reduce uncertainty (see Section 2.1), effective feedback mechanisms, and discussions to reach a wide consensus about what is desirable and acceptable (Dosi, Nelson, and Winter 2000; Davila, Epstein, and Shelton 2006). The efficiency of learning and feedback mechanisms is fundamental, because the number of options that can be tried is limited by coordination problems in the organization, the resources available for exploration of alternatives, and organizational learning capabilities (Levinthal 2000; Christensen and Raynor 2003).

Usually organizations find it difficult to learn and innovate. Successful routines hamper experimentation of potentially better approaches to problem solving, managers and employees are absorbed by routine activities, and mid-level managers filter ideas they think top managers will reject. To overcome these hurdles, it is recommended that specialized exploration units be established. These units should operate outside the organization's structure but develop strong links with the rest of the organization (Christensen, Anthony, and Roth 2004; Davila, Epstein, and Shelton 2006).

3.1.1 Profile of the Innovative Individual

Most innovation policies and programs are based on the implicit assumption that the entire target population has the capacity to innovate. This assumption is largely true for actors operating in dynamic markets, because competition pressures them to innovate to maintain or improve their competitive positions.[2] But in static markets, actors can survive without innovating, especially subsistence farmers, while other actors have to generate at least enough earnings to feed themselves. In other words, competition functions as a selection mechanism based on, among other factors, innovative capabilities.

Individual innovative capabilities stem from the specific organization of cognitive experiences. These experiences result in representations of reality, that is, how each individual sees, understands, and interprets his or her sur-

[2]In the complexity literature, this phenomenon is called the Red Queen effect. In Lewis Caroll's book *Through the Looking Glass,* Alice says to the Red Queen, "in my country one runs to arrive at where one wants to go." And the Queen responds, "Now, here, you see, it takes all the running you can do to keep in the same place. If you want to get somewhere else, you must run at least twice as fast as that!" (Carroll 1982, 104; Robson 2005).

roundings. The essence of individual innovation is originality in the individual's representation of the world. Intelligent people, in particular innovators, see, understand, and interpret their reality differently from the rest of the world. Their representations are general, categorical, conceptually rich, complex, and unique, making possible exceptional performance and attainments (Shavinina and Seeratan 2003).

Innovative individuals distinguish themselves by having general and specific abilities superior to the average (although not necessarily exceptional), being strongly committed to a project, and being creative (Renzulli 2003; Shavinina and Seeratan 2003). Innovative talent depends on the interaction among these three factors and not on possessing exceptional capabilities in one or two of them. Two other factors that affect the manifestation of innovative capabilities are personality and the environment (Bailey and Ford 2003; Georgsdottir, Lubart, and Get 2003).

General abilities are capabilities to process information, integrate experiences, and have abstract thoughts, whereas specific abilities are the capabilities to acquire specialized knowledge or skills. Outstanding among the general abilities are academic capabilities, whereas scientific, artistic, and sporting skills are important examples of the specific abilities (Shavinina and Seeratan 2003). It has been shown that intellectually talented people can be distinguished by a knowledge stock that is adequate, sophisticated, and well structured, which they can access easily. Possession of specialized knowledge or an innovative profile is not necessarily linked to formal education but more to learning processes (see Section 3.1.2).

Task commitment is also defined as perseverance, specialized practice, self-confidence, and the ability to carry out major tasks independently. There is no universally accepted definition of creativity, but in general, it is recognized that creativity is related to the ability to think things that nobody else has thought.

Although these abilities cannot be measured quantitatively, it is estimated that innovators fall into the upper 20 percent of the distribution in each one of these factors (Renzulli 2003); thus less that 1 percent of the population will simultaneously and adequately possess the three attributes necessary for innovation.

"A person is not always innovative or never innovative. Innovation occurs in certain people (not all people), at certain times (not all the time), and under certain circumstances" (Renzulli 2003, 83). Some people have only one innovative episode in their lives; others maintain their innovative capabilities for long periods.[3] No one is an innovator at all times or never, as opposed to other abilities that manifest themselves more or less permanently (for example, sporting or artistic abilities). Furthermore, those who are consistently more creative have had noncreative periods after long and intense manifestations of their talents (Renzulli 2003).

Innovative capabilities can be developed in some people if they are adequately stimulated and trained for those activities for which they have talent and they interact appropriately with the environment (Renzulli 2003; Vandervert 2003). However, individual responses to stimuli have great variability; furthermore, it is impossible to predict which individual will respond favorably to given types of stimuli. Consequently, in the design of programs to develop innovative capabilities it is preferable to include rather than exclude those who have potential to develop innovative ideas or products (Renzulli 2003).

The literature reviewed above only focuses on innate and acquired idiosyncratic factors. In fact, the expression of these factors depends on

[3]For example, Giusepe Tomasi de Lampedusa wrote a single novel that became a classic of literature, whereas Thomas Edison obtained more than 2,000 patents.

other variables, such as wealth, luck, experience, culture, and gender. For example, in a study of innovative African farmers, Reij and Waters-Bayer (2001) found that in most cases they (1) were men (reflecting differences in asset ownership and cultural discrimination); (2) had strong personalities; (3) were relatively old and had great experience; (4) were relatively rich (which enabled them to experiment more); (5) had contacts with other regions, which were a source of technological information; (6) were full-time farmers; (7) showed no correlation between creativity and formal education; and (8) tended to develop integrated agricultural systems.

3.1.2 Learning

Learning is defined as the process by which people and organizations create knowledge and acquire capabilities. It is a complex process based on repetition, experimentation, and selection, which enables improved and faster performance of tasks and the identification of new opportunities (Dodgson 1993). Learning processes have a gradual, cumulative, systemic, and idiosyncratic character.

Knowledge is the fruit of the learning process, and the application of knowledge is a feedback to the process. Learning has been recognized as an essential component of human capital, but it is a special form of capital, because it increases with use and depreciates if it is not applied (OECD 1996).

Learning takes place basically at three levels: individual, organizational, and network. Simon (1996) points out that all learning begins at the individual level—thus an organization learns in two ways: (1) through the learning of its members, or (2) through incorporating new members who have knowledge that the organization does not possess. Thus the definition of learning, and therefore of knowledge, is in the first instance an individual affair. But this classification is insufficient, as it does not account for knowledge developed jointly by several individuals.

Individual learning is also a social phenomenon: what an individual learns largely depends on what is known by other members of the organization and what the organization allows in terms of experimentation (Von Krogh, Ichijo, and Nonaka 2000; Bailey and Ford 2003). In general, organizations do not allow for innovations, because they weaken routines that have yielded positive results in the past and are still considered useful (Bailey and Ford 2003; Davila, Epstein, and Shelton 2006). Collective knowledge depends on routines and conventions developed (often unconsciously) and accepted by members of the organization. In turn, through interactions, individuals define new operational routines; that is, routines are an expression of collective learning (Dosi, Nelson, and Winter 2000). Therefore, although organizational learning occurs through individuals, it is not the sum of the individual learning of the organization's members (Dodgson 1993).

Organizational capabilities are a combination of elements sometimes explicit and articulated and sometimes tacit and subconscious (Dosi, Nelson, and Winter 2000). For this reason, two important components of organizational learning are (1) being able to share knowledge with different areas of the organization and with new members of the development team and (2) making tacit knowledge explicit so as to be able to reflect on it and transmit it across time, especially after the owners of the tacit knowledge leave the organization.

Collective learning occurs not only through individuals' imitation, like the master–apprentice relationship, but also as a result of the combination of individual efforts to understand complex problems. Learning requires developing shared codes of communication and search procedures. In organizations, routines are the manifestation of these codes and procedures (Teece, Pisano, and Shuen 1997). Organizations often are not aware of everything they know, making the construction of an institutional memory a key element of organizational learning.

Knowledge is complex, diverse, and volatile. In addition, its availability has grown exponentially in recent decades. For these reasons, companies increasingly need to cooperate and share information and innovative capabilities; that is, they need to integrate into networks. Innovation is increasingly the result of cooperation. Networks allow interchange of knowledge, abilities, and resources among different actors and institutions that make up systems at the local, regional, national, and international levels (Powell and Grodal 2005; Davila, Epstein, and Shelton 2006).

Lundvall (1992) highlights the interactive character of learning, which depends on the formation of networks made up of different actors from the innovation system, including companies, research centers, farmers, suppliers, and users. The effectiveness of these networks depends on flows of knowledge, the mechanisms used by agents and institutions to learn collectively, and the acquisition of capabilities through interaction. Strengthening the mechanisms by which learning and distribution of knowledge occur and encouraging the formation of networks have become key issues for decisionmakers.

A robust learning strategy needs to include various dimensions. First, it should simultaneously include both operational and strategic components. Second, it should define the goals of the learning, that is, what the organization needs to learn, including technological and organizational elements. Third, it should identify sources of learning and knowledge (internal, external, domestic, or foreign). Fourth, it should specify who learns (individuals, leaders, the organization, or networks). Fifth, it should identify the localization of the knowledge (that is, in which areas of the organization the knowledge will be produced). Sixth, it should identify the learning mechanism, in other words, the means or activities through which the actors learn (Vera-Cruz 2004).

Learning can occur through a number of activities, including production, research and development, reverse engineering, analysis of competitors' innovation portfolios (including the product basket and research and development projects), visits to other firms, team work, monitoring internal work, strategic alliances (with other firms and/or research institutions), and licensing.

In recent years ignorance management (that is, knowing what is not known) has been recognized as an important component of knowledge management (Root-Bernstein 2003; Davila, Epstein, and Shelton 2006).

3.1.3 Codified and Tacit Knowledge

Knowledge is a key component of capabilities. Knowledge is a concept related to data and information but is broader than these. Davenport and Prusak (1998) define knowledge as an evolving mix of structured experiences, values, contextualized information, and expert ideas, which help the organization to evaluate and incorporate new experiences and information. Knowledge originates and operates in the minds of those who know. In organizations it is frequently incorporated in organizational routines, processes, and rules. Documents and depositaries are not knowledge, they are just information—they become collective knowledge when they define how things are done in the organization.

An organization's knowledge includes both tacit and codified elements that are usually dispersed in different areas. Codified knowledge is articulated in formal language; that is, it is expressed with words, numbers, and drawings, including grammatical declarations, mathematical expressions, specifications, and manuals.[4] Tacit knowledge is difficult to articulate

[4]The concepts of explicit and codified knowledge are often used interchangeably. Nonaka and Takeuchi (1995) use the concept of explicit knowledge, but to preserve consistency with the concepts used in this study, codified knowledge is used in the remainder of this report.

through formal language; it is personal and accumulated in the course of individual and group experiences and is transmitted through practice. To be communicated and shared within the organization, it must be codified in some way; in other words, it must be converted into words or numbers that other people can understand (Dutrénit 2001, 2002; Nonaka and Takeuchi 1995). To use codified knowledge, it is essential to have a significant stock of tacit knowledge, for example, anyone can read a car repair manual, but only an experienced mechanic can repair the car adequately.

Knowledge can be generated from internal and external sources of information. The ability to use external information in productive and commercial activities is critical for the development and maintenance of innovative capabilities. Cohen and Levinthal (1990) called this ability absorptive capability. It depends on external factors (for example, economic stability, the competitiveness of markets, and relations between research institutions and organizations) and internal factors (including the company's culture, resources invested in the search for and adaptation of external information, the quality of its personnel, and mechanisms for socializing knowledge).

To exploit external information, among other factors, organizations must have accumulated some level of knowledge related to the new information. Making the new external information intelligible (for the organization) depends on the ability to link new and existing concepts. Organizations with stronger absorptive capabilities are more able to exploit emerging opportunities, whereas those with weaker absorptive capabilities tend to be more reactive.

An organization's absorptive capabilities are a function of individual and collective capabilities. The latter depend not only on interactions with the external environment but also on the organization's internal interfaces. Shared language and symbols facilitate the communication and effective interaction among different and diverse structures of knowledge. Acceptance of frequent revisions of operating routines and of searches for ways to improve operations (or to identify new areas where the organization can operate) also contributes to developing collective learning. Thus the organizational culture influences the development of absorptive capabilities (see Sections 3.1.1 and 3.1.2).

An organization's absorptive capabilities depend on the effort and resources that the organization has invested in them. In this sense, creation of absorptive capabilities is a deliberate process in which the intensity of effort is critical.

3.2 Organizational Culture

The role of organizational culture in shaping behavior has been receiving increasing attention. Organizational culture is a series of basic assumptions, values, and beliefs that were largely established during the organization's early life. These elements form a sort of organizational personality that emerges slowly and is hard to change but is not immutable. Culture affects learning and governance; some cultures facilitate individual learning, whereas others foster collective learning. Some emphasize vertical and centralized management, and others promote horizontal decisionmaking (Vera-Cruz 2004).

3.2.1 The Concept of Culture

Several definitions of organizational culture have been presented in the specialized literature. One of the most widely accepted states that culture is a pattern of basic assumptions and beliefs (not values or behaviors) invented, discovered, or developed by a group in the process of learning how to address the problems of external adaptation and internal integration (Schein 1984, 1991). This pattern has functioned well enough to be considered valid and, as such, is to be taught to new members of the organization as the correct way of perceiving, thinking, and feeling in relation to the problems of adaptation and integration. These basic assumptions and beliefs operate unconsciously and define an organization's view of itself and its environment.

This definition highlights the evolutionary forces that influence an organizational culture, especially the means by which it is learned, transmitted, and changed. Likewise, it emphasizes that a culture is rooted in the organization's need to solve problems associated with external circumstances that it had to face in its beginnings (external adaptation) plus the mechanisms used to integrate workers and cultivate their loyalty to the organization (internal integration). Finally, the definition stresses that construction of a culture is a protracted process that involves the repeated use of specific and successful approaches for solving similar problems.

3.2.2 Levels of Organizational Culture

Schein (1996) argued that a culture has three clearly differentiated levels. The first one corresponds to the "visible artifacts," that is, the physical and social surroundings constructed by the organization and the visible organizational processes and structures. They include the architecture, technology, office distribution, audible or visible patterns of behavior, and printed documents (for example, statutes, employee orientation materials, and reports). These data are readily available but difficult to interpret. This level describes how the organization builds its environment and what the patterns of behavior are, but not why they are what they are. Understanding why an organization behaves in a particular way requires jumping to the second level: the values that govern behavior.

Values refer to the ideal of what should be, and as such are difficult to apprehend directly. Values govern behavior and consist of strategies, goals, philosophies, and manifest justifications. They can be inferred from interviews with key members of the organization and the analysis of the visible artifacts. Nevertheless, they often are a manifestation of what a culture believes "should be" and not what the actual culture is. Although values could be present in the organization's unwritten code, that does not mean that they are applied in practice.

The third level includes the basic assumptions and refers to the organization's essential themes, such as human nature (good, bad, or neutral, and whether it can be improved), the nature of human activity (active, passive, or self-determined), the nature of human relations (cooperative or competitive, individualist or collaborative), the vision of reality (ways of accepting something as true, based on facts or the opinions of influential people), and the relationship with the surrounding world (to what degree the organization considers it can influence its environment). Fundamental assumptions are firm beliefs, perceptions, thoughts, and feelings taken as given; they are invisible and preconscious. They are very effective in explaining an organization's behavior. The basic assumptions are the essence of the organizational culture, whereas the values and behaviors are visible manifestations of this essence (Schein 1991).

3.2.3 Types of Culture

Whereas Schein (1991) seeks to understand the process by which an organizational culture is generated, Handy (1995) describes the basic types of cultures and pairs each type with a Greek god. He points out that "each culture . . . or each god works on quite different assumptions about the basis of power and influence, about what motivates people, how they think and learn, how things can be changed. These assumptions result in quite different styles of management, structures, procedures and reward systems" (Handy 1995, 11). These types of cultures bring with them different styles of administration, structures, procedures, and reward systems. Handy argues that each organizational culture is made up of different values and shared interests, resulting in specific ways of doing things and of evaluating what will function and what will not.

The culture types identified by Handy relevant for the study of the PFs are:

1. *Power, Club, or Zeus Culture.* This type is oriented toward activity. It is focused on the construction of useful relations to gain influence through informal channels. Although there may be a division of tasks based on function or products, true power is concentrated at the highest decision level and is distributed from the center to the periphery. The power of individuals in the organization depends on their relationships with the top decisionmakers and is based on empathy and trust. As a consequence, the opinions of influential people weigh more than those in technical positions. Outside the circle of power, relations are established through the hierarchy.
2. *Role or Apollo Culture.* This culture is based on the definition of specific roles and not on personalities. It emphasizes hierarchies, rules, clearly defined functions, and other structures that help people effectively do their assigned jobs. Individuals are important, but only as part of the organization.
3. *Task or Athena Culture.* This culture is focused on work activities. It appears in interdisciplinary groups organized around specific problems. There is an assumption of equality, where every opinion is important, and power is based on knowledge. The structure is quite decentralized and integrated into networks. Nevertheless, as different units have different specific responsibilities within a general strategy, this culture still preserves some formal elements. This structure is the one that most facilitates organizational learning.

Handy (1995) pointed out that many organizations were created as Club Cultures based on their founder's personality and, as the organization matured, they evolved into other forms. Most organizations contain aspects of several types of cultures; thus they have a mix of management styles. This mix does not lack contradictions; for example, a dominant personality (Zeus) in a horizontal (Athena) culture can generate enormous contradictions and discourage people from contributing to the organization, because the typical Zeus administrator is seen by Athenians as countercultural. Handy argues that each culture is good for something. No culture or mix of cultures is bad or erroneous in itself; it can be adequate in certain circumstances and inadequate in others.

Cultures resist change. Each culture generates great loyalty and unity on the part of its followers: those who are accustomed to one culture have difficulty in accepting the principles of another. Thus it is hard to change cultures, and many organizational problems result from attempting to continue doing things as usual, rejecting cultural changes when they are needed (Handy 1995; Christensen and Raynor 2003; Davila, Epstein, and Shelton 2006).

3.2.4 How an Organizational Culture Is Built

A culture's elements can be defined as learned solutions to problems (Schein 1996). In general, there are two types of situations that foment learning: (1) positive solutions to problems that generate self-reinforcing processes when the solutions work and (2) anxiety aversion, which generates self-reinforcing processes when solutions help avoid anxiety. An organization may have cultural elements based on both types of situations. The first case occurs when, to find positive solutions to problems, a group tries different alternatives until they find one that works. The repeated solution to the same problem through a similar response legitimates that mechanism as the correct way of solving the problem. In the second case, when a response that effectively avoids anxiety is found, the group repeats that response without checking whether the cause of the anxiety has disappeared.

An organization's founders shape its culture; they incorporate it and transmit it. Later, the organization's leaders (usually the top management) play a key role in transforming the culture (Vera-Cruz 2004; Davila, Epstein, and Shelton 2006). A culture depends on the group's experience; thus the process of culture formation is associated with the emergence of a group's identity.

Several external and internal factors drive the process of culture creation. External factors are related to the leaders' and group's perceptions regarding the external context and how to survive in it. Internal factors are associated to their perceptions on how to organize relations among members to enable survival in a given context—the relations should combine effective performance with internal comfort.

Problems of external adaptation and survival are associated with

- mission and strategy: obtaining a shared vision of the organization's mission, objectives derived from that mission, primary activities, and manifest and latent functions;
- means: developing a consensus about the means to be used to attain the objectives, such as organizational structure, division of labor, and system of authority; and
- incentive structure and metrics: developing a consensus on the criteria used to evaluate to what extent the group has attained its objectives (for example, on an information and control system); similarly, developing a consensus on appropriate remedial strategies in the case of nonattainment of objectives.

The problems associated with internal integration are

- developing a common language and conceptual categories, so that members can communicate and mutually understand each other;
- defining group borders and criteria on who can be a member and who cannot, and criteria to determine membership;
- defining the criteria and rules by which a person acquires, keeps, and loses power;
- setting what is rewarded with status and power and what is penalized; and
- establishing ideology and "religion," that is, the way events are interpreted.

3.3 Governance

Governance concerns all the rules, procedures, and practices affecting how power is exercised in an organization. The governance function of the organization is understood as a space where strategic decisions are taken, especially those relating to global resource allocation, and where the inevitable crises of any complex organization are faced and overcome.[5] Acknowledging that organizations establish a web of internal and external social interactions, the new organizational proposals are based on models of inter- and intra-organizational collaboration (Crozier 1994; Savall and Zardet 1995). In other words, organizations are seen as social spaces for collective action. The latter depend on the mental models of the individual as well as on organizational practices—especially on the work environment and the incentive structure. The key to an organization's success lies in convincing individuals that collective interests are more important than individual and group values, and in maintaining the governance to ensure the organization's survival through control of results and elimination of conflicts.

[5] In a complex organization, resource-allocation decisions are made continuously and at all levels. Low-level employees decide what actions to prioritize (for example, which clients to visit), middle managers select the options they submit to the senior management (eliminating the options they believe will be rejected by their supervisors), and executives make decisions partly in response to the feedback they receive from the employees (Christensen and Raynor 2003).

The new organizational models break with the traditional vision that sees organizations as closed, static systems isolated from their environments: the new vision recognizes organizations as complex systems, in which the organization functions as a space where diverse groups with different goals and interests coexist (Etkin 2000). This diversity can cause the organization to stray from its declared or foundational goals. Additionally, the organization is a system formed by a fuzzy set of plans, projects, and activities structured by ambiguous relations and contradictory processes. In this set, forces that push for change and to maintain the status quo coexist, so that the organization is a self-organizing system (see Section 2.1.1).

In the analysis of organizations, it is important to merge governance approaches with the complexity perspective to analyze such themes as legitimacy, decisionmaking, political participation, management representation, and the dimension of power. Analysis of the governance of organizations encompasses three dimensions: structure (distribution of functions and coordination), process (communication–coordination–harmonization, leadership, learning policies, and operative processes), and strategic axis (mission, shared vision, and strategic lines and plans of action).

3.3.1 Structure

An organization's structure has been defined as all the means used to divide the work into separate tasks and ensure their coordination (Mintzberg 1999). The structure defines authority, jobs, functions, responsibilities, and communication links among the organization's members. Mintzberg (1999) identified different organizational configurations, each corresponding to different coordination mechanisms:

- *Mutual adjustment.* Work coordination is carried out by the simple process of informal communication.
- *Direct supervision.* Coordination is carried out by one person who is in charge of employees.
- *Standardization of work process.* Coordination is carried out through fulfilling the functions established in the procedures manual.
- *Standardization of results.* Coordination is defined on the basis of the results that must be obtained.
- *Standardization of qualifications.* Coordination of different types of work is carried out based on the know-how of those who do the work.
- *Standardization of rules.* Workers carry out their functions based on the rules that are one component of the organizational culture.

The existence of a number of alternative structures reflects the complex social nature of the organization, that is, an organism plagued by potential conflicts among actors and by the dysfunctions the structures may provoke. To maintain unity, organizations must permanently coordinate areas and spaces with divergent modes of thought, resources, and priorities. The structural model thus defines the degree of centralization of responsibilities and the degree of autonomy and cohesion of tasks that condition the development of new skills necessary for survival in a changing world. To develop an effective structure, an organization's government should seek out configurations that (1) are effective and congruent with the goals and resources, (2) identify limitations of the hegemonic model that allowed survival in the past, and (3) allow the emergence of alternatives.[6]

In an alternative classification of organizational structures, Gibson, Ivancevich, and Donnelly (1996) described two basic models. The *organizational model* refers to a mechanistic model that emphasizes achieving high

[6]The organization's government consists of those people or groups who make the strategic decisions.

levels of production and efficiency through the intensive use of rules and procedures, centralization of authority, and strict work specialization. In contrast, the *organic model* emphasizes the importance of achieving high levels of adaptability and development through the limited use of rules and procedures, decentralization of authority, and relatively low levels of specialization. Theories of contingent design are based on these two models. These theories hold that the organization's structure should fit the needs of a given situation, including issues related to technology, environmental uncertainties, and strategy.

In the mechanistic model of the organization, decisionmaking is concentrated at the pinnacle of the organization. The organization is designed to give the most authority and responsibility to the highest position in the hierarchical structure, because the person occupying this position is supposed to possess greater ability and knowledge than the lower ranks. Such a management model is based on a generalized distrust of the performance of the lower ranks, justifying a tight control over the organization's implicit knowledge. It is now recognized that these models only create the illusion of control, because all employees have a certain amount of power, even if it is only the power to obstruct the decisions of their superiors (Olson and Eoyang 2001).

In contrast, the organic model is based on decentralized and flexible structures built on confidence. This model starts from the idea that every actor has the ability to make decisions according to his or her position in the hierarchy (Olson and Eoyang 2001; Christensen and Raynor 2003). Thus the organization recognizes the accumulated experience of all actors and the possibilities that are opened with the active participation of all members of the organization. The success of the decentralization process depends on all actors internalizing the organization's vision and mission.

3.3.2 Organizational Processes

An organization takes shape through its processes, that is, recurrent activities that use resources. The study of processes allows visualization of the decisionmaking mechanisms (beyond what figures in organizational charts or operations manuals) and behaviors that result in specific outcomes, successful or otherwise. The principal processes are communication–coordination–harmonization, leadership, politics, learning, and the organization's operational processes.

Communication–coordination–harmonization process. Subjects in an organization interact across space and time, pursuing their individual and collective interests. These subjects belong to different social groups (for example, social, professional, occupational, or educational classes) and have diverse experiences, which can lead them into idiosyncratic behaviors.

Because the capacity for action, decisionmaking, the establishment of strategies, and the organization's shared vision depend on the abilities and resources acquired by its members, the organization must generate adequate conditions for collective action. Horizontal communication processes help to generate internal consensus through guided discussion, easing the transition from an individual to a shared organizational vision. These processes generate collective or organizational knowledge.

Organizations often confuse information with communication, because internal reports and memos automatically foment the creation of shared meanings and values, which, in turn, generate a sense of belonging. On many occasions, however, concise memos lead to a sense of dissonance or separation from the organization. Thus, instead of strengthening the collective identity, these brief communications provoke a distancing between actors, especially between those who have authority and those who do not.

The organization's internal communication process is present in all social interchanges. Communication affects the construction, comprehension, and influence of multiple meanings that are transmitted in all social spaces. Communications form an indispensable part of cultural processes, which is why it is important to put into place operational, functional, and stimulating information systems. Such systems motivate action through a continuous negotiation process (Savall and Zardet 1995).

Leadership: Managing, deciding, influencing, and motivating processes. Most experts on human behavior have tried to correlate leadership characteristics with performance. The consensus today is that a modern leader must develop leadership with a sense of duty; in other words, she or he must project into the interior of the organization a commitment that functions as a reference for the transformation of organizational structures. The leader has to create a symbol of leadership that serves as an example to promote a more committed and responsible participation that responds more to conviction than to coercion exercised by authority and induces members to contribute their knowledge, abilities, and experience toward common goals.

An effective leader motivates, shows personal interest, generates an environment of confidence, develops the capacity for self-control, exercises effective leadership that facilitates rapid decisionmaking on the part of collaborators, and awards opportune recognition for work well done. In addition, he or she creates a sociable work environment, is a communicator, and demonstrates willingness and ability to listen and keep an open dialogue to share information. He or she explains and transmits to the team a clear vision of the present situation and possible future outlook (Crozier 1994). The leader should also encourage and facilitate the personal and professional development of the team members and participate in learning, training, and development processes while ensuring that the continuous improvement process in his or her area of responsibility is appropriately documented (Luhmann 1997). In his study of the role of the leader in the promotion and guidance of the administrative process, Mintzberg (1999) found that leaders do not simply plan, organize, direct, and control; they also fulfill three fundamental roles: interpersonal, informative, and decisive (Table 3.1).

Political processes in the organization. The political approach sees the organization as a space in which different groups (with their own goals) coexist and make formal or tacit agreements on matters relevant to the organization, including the form of governance, the rules of the game (normative framework), resource distribution criteria, and negotiations with other institutions. The political axis of the organization seeks areas of coincidence and establishes means of resolving the conflicts that stem from a diversity of goals.

Political analysis attempts to discover and negotiate hidden aspects, including declared interests that guide decisions and exist in parallel with the official objectives, forces that operate on the margins of or outside formal structures, and negotiations among interest groups that do not interact within the traditional channels of communication (Etkin 2000). The mechanisms used in political dynamics include power plays, alliances, and coalitions, negotiation processes, public consultations, and assemblies called to defend rights or promote changes in labor relations.

Power: Sources and mechanisms; formal and informal relations. The concept of power is of vital importance, not only because every organization has some kind of hierarchical structure, but also because the organization establishes power relations with other organizations or political institutions. The definition of power in its most gen-

Table 3.1 Role of the leader

Interpersonal	
Figurehead	Symbolic head, needed for numerous legal and social duties
Leader	Responsible for leading and motivating subordinates
Connector	Maintains a network of external contacts for favors and information
Informative	
Observer	Receives a wide variety of information; functions as an internal–external node of information for the organization
Disseminator	Transmits information received from subordinates or external sources to the organization's members
Spokesperson	Transmits information to external entities on the organization's plans, policies, actions, and results; serves as an expert in the organization's industry
Decisionmaker	
Entrepreneur	Searches out and initiates projects that lead to organizational change
Obstacle manager	Responsible for corrective action when the organization confronts significant or unexpected obstacles
Resource distributor	Makes or approves significant organizational decisions
Negotiator	Responsible for representing the organization in major negotiations

Source: Robbins (1999).

eral sense refers to the capacity or possibility of producing effects (Bobbio and Matteucci 1988). As a social phenomenon, power is the relation among individuals. Weber (1984) defines power as the possibility of imposing one's own will in a social relation, even in the face of resistance and whatever the basis of that relation is. Power relations can be based on the control of resources valued by a group (Adams 1993), social conventions, or psychological factors (subjection of a weak character by a strong one).

Power confers on its holders the capacity to induce a person or group of people to work toward a desired objective. But power is not absolute; all actors possess a degree of power that allows them to choose a strategy to reach their individual or group objectives (Crozier and Friedberg 1977).

Panebianco (1990) considers power as a relation of reciprocal interchange in which the terms of interchange favor one of the parties; in other words, it is an asymmetric relation. Power relations are established among leaders and/or elites on the one hand, and among followers on the other. In these relations, the followers support the leader if they receive something in return. There are two types of power relations: horizontal (among the power elite or among followers) and vertical (between elite and followers). Both types of relations are found in the interior and the exterior of the organization. However, in the study of power structures it is not sufficient to simply be acquainted with actors and the power relations they establish. It is also necessary to understand the content of the interchange, where it is most concentrated, the content of the negotiations, the set of incentives (collective and/or selective) that elicit participation, and the benefits obtained by the actors in the negotiation.

The sources of power are the elements that confer the ability to influence other actors. Five types of power, based on their source, have been identified: coercive, reward-based, legitimate, expert, and referential (Robbins 1999):

- *Coercive power.* Based on fear of the consequences of not obeying, it rests on either the threat or the actual application of sanctions.
- *Reward-based power.* People obey to obtain benefits, and power rests on the ability to distribute rewards or benefits.

- *Legitimate power.* Each position in an organization's formal hierarchy has a certain amount of influence attached to it, and power requires that the organization's members accept the authority derived from the position.
- *Expert power.* Influence is based on experience, skill, or knowledge. This is one of the most powerful sources of influence, because it is based on the respect that recognition brings.
- *Referential power.* Based on the identification with someone with desirable resources or personal characteristics, it rests on admiration and the desire to emulate the admired person.

Alternatively, Foucault (1978) considered three types of power:

- *Economic power* is related to asset ownership (in particular capital) and the purchasing power such ownership confers.
- *Ideological power* is related to the ability to impose a world view.
- *Government power* is that conferred by a community on an individual or a small group to allocate the goods produced by the community, according to agreed rules.

Operating process: The organization's raison d'etre. Organizational processes are a set of interrelated recurrent activities that serve to maintain and develop the organization and to guide its interactions with the external environment. Changes that occur in processes in response to external developments are aimed at guaranteeing efficient performance of the organization's basic operations and its survival. Actual processes are the result of purposeful design, chance, and past experiences; as such, they are closely related to routines (Dosi, Nelson, and Winter 2000).

Governing the organization. The function of government is to promote a creative and proactive attitude for searching for long-term opportunities, given the external and internal contexts. A manager should challenge existing routines without fostering excessive change. A leader manages the organization's relations with the external world and, in this context, steers the organization's evaluation of expansion opportunities, new demands, changes in its environment, and establishment of strategic alliances with other organizations (Christensen and Raynor 2003). The leader helps to mediate and resolve internal conflicts. Success of the government function thus depends on leadership that promotes the emergence of a set of opinions and shared values. Another key facet of the government function is the definition of strategic plans to

- act on the organization's external environment; for example, success in the market (these objectives constitute the organization's visible, organized, and planned superstructure) and
- act on the organization's internal environment to reduce dysfunctions (these actions are directed at six areas: working conditions, the organization of work, communication–coordination–harmonization, time use, integrated training, and strategy implementation; the internal environment nearly always corresponds to the hidden, unplanned, badly organized part of the strategy) (Savall and Zardet 1995).

The role of a strategy is to create new capabilities that enable advances in the organizational learning process at both the individual and collective levels. Savall and Zardet (1995) mention four fundamental attributes of the strategy:

1. Human potential is essential in the search for new strategies.
2. An effective strategy results from the participation of all organizational actors in a process that is negotiated through the hierarchy.

3. An organization's strategy is an attempt to conciliate, partially but significantly and effectively, the individual strategies of internal and external actors as well as those of actors in different categories.
4. A strategy simultaneously defines medium-term goals and the means to reach them. The concerted strategic planning process legitimates the strategic goals and action plans, leading to more effective implementation.

The indiscriminate use of strategic planning has been criticized in recent years (Olson and Eoyang 2001; Christensen and Raynor 2003). In emergent processes (new technologies or markets) the information to carry out a deep analysis of the situation in which the organization is going to operate simply does not exist. In these cases, strategic planning is too structured, and more flexible planning methods should be used (see Section 2.1.2).

3.4 Innovative Capabilities and Poverty Alleviation

Development depends on the ability of a number of actors to develop individual and collective innovative capabilities. This ability is particularly important for poverty alleviation, because poor rural households have to find new strategies to integrate into globalized economies (Berdegué 2005). Although many authors have analyzed innovation processes that involve small farmers, and even more agencies have implemented programs to help these farmers to innovate, few authors have systematically researched how to build the institutional capabilities that agencies need to strengthen their impact on poverty. Section 3.4.1 briefly surveys the literature on innovation in small farms, focusing on recommendations to build institutional capabilities to foster pro-poor innovations. Section 3.4.2 explores how the skewed distribution of innovation capabilities influences the impact of poverty-alleviation policies.

3.4.1 What Is Known about Agricultural Innovation Policies and Poverty Alleviation?

The study of agricultural innovations and how they relate to poverty alleviation has evolved essentially from two sources: one that started with the Farming Systems Research approach in the 1970s and another that, since the late 1980s, has analyzed the organization of agricultural research and extension systems. In recent years, the analysis of agricultural innovation systems has exploded. The analysis of this literature, however, exceeds the scope of this report; a good literature survey can be found in Hirvonen (2007). Many publications on agricultural innovation systems are descriptive and provide general policy recommendations. This section reviews a few documents that contain concrete guidelines to foster pro-poor agricultural innovations.

This literature has influenced the activities of several organizations that work in poverty alleviation (including the World Bank, the International Fund for Agricultural Development [IFAD], Consultative Group on International Agricultural Research [CGIAR] centers, universities from developed countries, and a number of NGOs) that have explored how the innovation systems framework could help them to strengthen their impact. It has been difficult, though, to incorporate this framework into their working routines (Berdegué 2005; Rodrik 2006; Smit 2007). Two reasons explain this difficulty: on the one hand, organizations resist change (see Sections 3.1 and 3.2); on the other hand, because of the complexity of innovation processes, there are no clear recipes on how to make the innovation systems operative (World Bank 2006a).

Although the World Bank, the Inter-American Development Bank, and IFAD have explored the use of the innovation systems framework in their operations, these attempts have been isolated. Among the CGIAR centers that have explored how to use this framework, Bioversity, the International Potato Center (CIP), the International Center for Tropical Agriculture (CIAT), and the International Livestock Research Institute stand out; among the universities, Wageningen and the United Nations University–Merit have active research programs related to agricultural innovations. The Institute of Development Studies has published seminal works in Farming Systems Research, in particular, Chambers, Pacey, and Thrupp (1989).

Berdegué (2005) stresses that the diversity of rural poverty prevents applying one-size-fits-all recipes and suggests that IFAD could become a promoter of experimentation, financing new types of projects for which it should build the appropriate capacities. Hall, Clark, and Naik (2007) also recognize the importance of capacity building for developing pro-poor innovations.

World Bank (2006a) presents a conceptual framework to analyze agricultural innovation systems and an intervention framework to guide the design of interventions according to different stages of development of the innovation processes. Hall, Clark, and Naik (2007) and World Bank (2006a) both stress the importance of self-organization and emergent innovation processes. Biggs and Matsaert (2004), Ferris et al. (2006), and Hartwich et al. (2008) discuss several methods to build innovation networks, and Biggs (2005) presents a method to develop innovations by learning from positive experiences.

This report builds on these works and complexity theories to develop guidelines on how to foster pro-poor innovations. The report's main insight is that relying on emergent processes is effective only when variety and selection are managed appropriately—otherwise, too many resources are lost in random experimentation (see Section 2.2). To do this, appropriate capabilities must be built, as discussed not only in this chapter, but also in Chapter 7 and Appendix C.

3.4.2 How Should Uneven Innovative Capabilities Influence Poverty-Alleviation Policies?

Currently, poor rural households can follow three pathways to escape poverty: becoming commercial farmers, seeking off-farm employment, or migrating (World Bank 2007). The combination of paths each household chooses depends on its accumulated assets, including its human and social capital and its innovative capabilities. Innovation policies can contribute to poverty alleviation along the first two pathways. Supporting commercial agriculture creates employment and reduces food prices. Many instruments have been found to be effective in supporting commercial farmers (see, for example, Table 2.1). Because the focus of this section is on the role of innovation policies in supporting small farmers, instruments to support commercial agriculture will not be analyzed.

Several innovation policies can help small farmers become entrepreneurs, in particular (1) helping them to build their individual and collective innovative capabilities, (2) teaching them how to integrate into value chains and to search for and use commercial and technical information, (3) facilitating the diffusion of information and devising new extension methods, (4) building physical infrastructure, and (5) reducing market inefficiencies. Because the development of entrepreneurial capabilities is a complex problem (see below), it is impossible to know in advance which menu of policies will be effective at a particular juncture.

Finding the right combination of programs requires an evolutionary approach to policy-making (see Section 2.1). Several agents, including NGOs (for example, Technoserve), advanced research institutions (for example, Michigan State University), and international research centers (such as CIP and CIAT) have implemented programs to develop value chains and market niches for small farmers. Few, however, have had more than local success, an indication of the complexity of poverty alleviation. As the experience of the PFs clearly shows (see Section 6.5), in addition to a flexible approach to project management, the presence of individuals highly committed to poverty alleviation in charge of the experimentation is a crucial factor in project performance.

Innovation results from the interaction of motivation and capabilities. Motivation for economic innovations arises from commercial opportunities. To take advantage of these opportunities, innovators must have the individual and collective capabilities to develop a package that combines appropriate technologies with a business plan (Christensen, Anthony, and Roth 2004; Davila, Epstein, and Shelton 2006). As was explained in Section 3.1.1, it is highly probable that only a fraction of small farmers would have such capabilities. Two current processes provide indirect support to this assertion. First, although many projects have sought to facilitate the access of small farmers to high-value chains, very few cases have been documented in which small farmers remained important suppliers when these chains matured (see, for example, Holden, Shiferaw, and Pender 2004; World Bank 2005, 2007; Reardon and Flores 2006). In other words, many small farmers try but few succeed.

Second, remittances to rural households have soared in the past two decades. International remittances in 2006 were estimated at US$300 billion (IFAD 2008). There are no estimates of the remittances sent by domestic migrants, but undoubtedly they also contributed significantly to diversifying the income of rural households. It has been consistently found that remittances are mostly invested in education and health (that is, in human capital that can be used in off-farm employment) and housing, and only a small share is used in productive activities, especially agriculture (Davis et al. 2000; Vargas-Lundius 2004; López-Córdova and Olmedo 2006). In fact, it has been found that for many of these households, increasing agricultural productivity has become less relevant than expanding other sources of income (Davis et al. 2000; Barrett, Reardon, and Webb 2001; World Bank 2007). In short, agricultural production is not one of the main priorities of households that receive remittances.

The effectiveness of innovation programs can be enhanced by targeting farmers with strong innovative capabilities. This targeting, however, is complex, as attested by the literature on programs for gifted students (Reis and Renzulli 2003).[7] Devising targeting mechanisms requires exploration and effective learning routines (see Section 2.1.2). Few organizations that implement innovation programs, however, have developed learning capabilities; most are trapped into what has been called the "To Do Mode" (Smit 2007). The PFs, however, are investing actively to develop such capabilities (see Section 6.5 and Chapter 8).

Despite the rapid growth of commercial agriculture, most poor rural households still depend on traditional crops for their living (World Bank 2007). For these households, higher and more stable yields can increase income and reduce food insecurity (Meinzen-Dick et al. 2004). It is unlikely, however, that they will be able to escape poverty producing only traditional crops, even if they use the best available technologies. In some cases, the higher output of traditional crops has enabled poor farmers to start new income-generating activities (Ekboir, Boa, and Dankyi 2002; Meinzen-Dick et al. 2004), but it is these new activities that have allowed them to escape poverty. And the new activities a rural house-

[7]Identifying farmers with strong entrepreneurial capabilities among all small farmers is similar in nature to identifying gifted students among the general student population.

hold can start depend, again, on their assets and capabilities.

Traditionally it was assumed that formal research contributed to poverty alleviation by developing technologies that poor farmers could use. As explained above, however, the use of technical information depends on the farmers' absorptive capabilities. Therefore, research policies can help small farmers mainly by facilitating the researchers' integration into innovation networks, which can be achieved by promoting changes in the researchers' cultures and creating new incentives that do not emphasize peer-reviewed publications. As seen in Sections 6.4, 6.5, and Appendix C, the PFs are already exploring some of these incentives.

CHAPTER **4**

Concepts and Methods

Social processes can be analyzed as complex systems in which different actors interact through multiple channels that include positive and negative feedback loops. Additionally, these interactions and their intensities change over time, often in unpredictable ways (see Section 2.1.1). The study of such processes requires flexible approaches to discover the emergence of new actors and changes in the patterns of interaction. Section 4.1 presents the conceptual framework used in the empirical analysis. Section 4.2 reviews the role of quantitative and qualitative methods in the analysis of complex social processes, and Section 4.3 describes the methodology used in the empirical analysis.

4.1 Conceptual Framework

Chapters 2 and 3 presented a detailed review of relatively new concepts that have appeared in usually separate branches of the scientific literature. The detail given in these chapters is greater than needed in the present empirical analysis; the extra material was kept in this report, however, because it is a useful review of topics that are not always easy to access.

These concepts are the basis of the conceptual framework used in the institutional analysis of the PFs. This framework views organizations as complex organisms influenced by the interaction between motivation and capabilities. Motivation results from the opportunities for change: in private firms these are mainly market or technological opportunities; for civil society organizations, the main driver is a sense of duty and the importance of the socioeconomic problems being addressed. The capabilities determine how organizations respond to emerging opportunities. Capabilities are constructed through sustained investments and depend on a number of factors, including the organization's past, its culture, governance structures, the presence of innovative individuals, and the environment in which the organization operates. Each factor's influence on the organization's dynamics changes over time, so that variables that have a positive effect at a particular point in time can be detrimental later (Christensen and Raynor 2003; Davila, Epstein, and Shelton 2006).

The dynamics of organizational capabilities and their interaction with external factors can be explained by the properties of CASs. These systems evolve through the interactions among a large number of actors of different types, conditioned by the process's history, the socioeconomic environment in which they operate, and random events (Kauffman 1995). In the case of agricultural innovations, the actors may include farmers, private firms, researchers, NGOs, farmers' associations, financial institutions, and policymakers. This section discusses organizational innovations as complex processes, the nature of organizational capabilities (especially the role of learning, institutional cultures, and governance structures), and the role of networks in innovation processes.

4.1.1 Organizational Innovations as Complex Processes

Innovation processes are an example of CAS. Self-organization is one of the key features of CAS: properties that do not exist at smaller scales emerge at larger ones from the actions and interactions among different actors (Kauffman 1995; Axelrod and Cohen 1999; Crutchfield and Schuster 2003). For example, life results from a huge number of chemical reactions, but each reaction is not alive. The interaction between self-organization and random events is the main reason that CASs are uncertain and unpredictable.

The most important instrument to influence a CAS is the creation of variation combined with effective selection mechanisms. In natural systems, variation is random, and selection is based on reproductive efficiency. Human interventions operate on both mechanisms, modifying them in a directed fashion, as exemplified by plant breeding (see Box 2.2). The effectiveness of this method depends crucially on having appropriate selection mechanisms. Because of the system's complexity, though, there are no optimal selection mechanisms: appropriate solutions can only be found with a directed trial-and-error approach based on strong learning capabilities (Crutchfield and Schuster 2003).

Organizational strategies can be of two types: deliberate and emergent. Deliberate strategies are those consciously defined by the organization, whereas emergent strategies are those that result from the day-to-day, decentralized actions of all decisionmakers in the organization (for example, top and middle managers, sales people, and floor workers). These actions tend to be tactical and include, for example, resource allocations within divisions or deciding which clients should be prioritized. The accumulation of these operational decisions results in the organization's actual resource allocation, which may differ from the allocation decided in the deliberate strategies (Christensen, Anthony, and Roth 2004). Without strong leadership and effective learning mechanisms, emergent strategies dominate the evolution of organizations. In such cases, organizations cannot operate consciously to take advantage of emergent opportunities. The PFs evolved through a combination of emergent and deliberate strategies, whose relative importance changed on several occasions over time.

4.1.2 Organizational Innovative Capabilities

Three of the most important institutional features that influence organizational innovative capabilities are learning mechanisms (because they determine how fast new information can be developed and used), organizational culture (because it determines the often "unconscious" factors that influence innovation), and governance structure (because it defines the flexibility an organization has to organize its activities and resolve conflicts).

Organizational learning. Innovation processes are complex because if they could be planned in detail, they would be routines, not innovations. Innovations result not only from variation (trying new things) but also from directed selection (finding things that are better than those in use). Managing variation and selection depends on the innovators' ability to test new things until a solution to a problem or opportunity is found (Crutchfield and Schuster 2003). This ability, in turn, depends on the capability to integrate into innovation networks and the network's ability to facilitate the exchange of resources and information. In other words, innovations depend on the interaction between individual and collective (organizational) learning capabilities (Dosi, Nelson, and Winter 2000).

Individual abilities result from the combination of a person's innate talents and learned abilities (Renzulli 2003). In contrast, organizational capabilities result from the interaction among the organization's resources (individuals' abilities and fixed capital), processes (the mechanisms agreed on by the organization for doing things), and values (including the institutional culture and long-term goals). In new organizations, most capabilities reside in the resources, especially their people (including managers, technical support personnel, and line workers). The incorporation or departure of a key person can have a major influence on the probability of success. In time, the capabilities of successful organizations are transferred to processes and values (Christensen and Raynor 2003). Because of the crucial importance of individuals in the organizations' early life, this report explores how innovators interacted among themselves and reacted to external influences to create strong innovative capabilities and set the basis for the institutional culture.

Organizational capabilities must be built through sustained investments, strong leadership, and adequate selection of those responsible for new projects. It is also necessary that top managers install a vision of the changes to be introduced, conduct many trials to reduce uncertainty, develop effective feedback mechanisms, and foster discussions to reach a consensus about what is desirable and acceptable (Levinthal 2000; Crutchfield and Schuster 2003; Davila, Epstein, and Shelton 2006). The effectiveness of learning and feedback mechanisms is fundamental, because the number of options that can be tried is limited by coordination problems in the organization, the resources available for exploration of alternatives, and organizational capabilities (Dosi, Nelson, and Winter 2000). Learning is also important to define effective deliberate strategies.

Usually organizations find it difficult to learn and innovate. Successful routines hamper testing potentially better approaches to problem solving, and managers and employees are absorbed by routine activities (Dosi, Nelson, and Winter 2000; Christensen and Raynor 2003). This difficulty also affected the PFs at some stages of their development, but most of the times they were able to develop new innovative capabilities.

Learning takes place basically at three levels: individual, organizational, and network. Learning begins at the individual level, enabling people to process information to build conceptual frameworks. Individual learning is also a collective process, because what an individual learns largely depends on what is known by other members of the group and what the organization allows in terms of experimentation (Von Krogh, Ichijo, and Nonaka 2000; Bailey and Ford 2003). Organizational learning is the process whereby knowledge is created and distributed across the organization, communicated among its members, and integrated into the organization's operations (Dodgson 1993; Dosi, Nelson, and Winter 2000). Network learning involves several individuals and organizations creating shared knowledge and is a special case of organizational learning.

Learning can occur through several mechanisms; for the analysis of the PFs, the most important mechanisms are learning by doing, from hiring key individuals, from interacting with partners, and from information searches.

Organizational culture. An organizational culture is a series of basic assumptions, values, and beliefs largely developed during an organization's early life. Organizational cultures determine, to a great extent, what is acceptable within an organization (Schein 1996). Some cultures facilitate individual learning, whereas others foment organizational learning. Some emphasize vertical and centralized management, whereas others promote horizontal decisionmaking. Two of the basic types of cultures identified in the literature are used in the analysis of the organizational culture of the PFs (Handy 1995; see also Section 3.2.3):

- Power, Club, or Zeus culture is oriented toward activity. The power of the individu-

als within the organization depends on their relationship with the highest level of management and is based on empathy and trust.
- Task or Athena culture is focused on work activities, integrated by interdisciplinary groups organized around solving specific problems. Power is based on knowledge in a quite decentralized structure.

No type of culture is absolutely better than the others; some types are better in particular situations and can be a liability in others or as the organization matures. The cultures of the PFs were developed by two distinct groups, the presidents and the managers, and their interaction was a major influence on how the PFs evolved and learned.

Governance structure. The analysis of governance of the PFs focuses on three dimensions: (1) structure (distribution of functions and coordination), (2) process (communication–coordination–harmonization, leadership, learning policies, and operative processes), and (3) the strategic axis (mission, shared vision, and strategic lines and plans of action). The structure defines authority, jobs, functions, responsibilities, and nexuses of communication among the organization's parts; processes are recurrent activities that use organizational resources; and the strategic axis defines the objectives and action plans.

Whereas organizational cultures change very slowly and are mainly an emergent property of the organization's evolution, governance can change relatively rapidly and as a result of deliberate action (Mintzberg 1999).

4.1.3 Innovation Networks

Individual actors (including firms) generally do not possess all the resources they need to innovate; therefore, actors integrate into networks that facilitate the interchange of knowledge, abilities, and resources (Powell and Grodal 2005). The evolution of innovation networks is determined by the changing relationships among agents, technologies, markets, the formal and informal rules that regulate people's behavior, and the complexity and maturity of the innovations. For relatively simple or mature innovations and markets, networks are rather lax; members often relate formally or are mediated by markets because each agent has a relatively good understanding of the needs of partners and customers, and of the technological and market opportunities. In contrast, in the case of complex or new innovations, partners have to interact informally, often, and intensely in order to overcome emergent hurdles. Because most of the innovations are radical and/or the markets change rapidly, these networks face great technical and commercial uncertainties, which prevent effective contracting: successful collaborations are based on trust. In fact, the degree of formality varies along a continuum that goes from completely formal to absolutely informal. Moreover, the characterization of a market or innovation as simple and formal or complex and informal can change, reflecting new technologies or commercial opportunities: as the markets evolve, membership in the network changes, reflecting changes in the agents' objectives and emerging technological challenges (Rycroft and Kash 1999).

A catalyzing agent is one of the most important factors in the emergence and consolidation of innovation networks. This agent induces partners to invest time and resources in the network. Once the network is consolidated, the importance of the catalyzing agent diminishes, because other actors are more willing to participate when the benefits of participation become clearer and the interaction rules are known to all partners.

4.2 Quantitative and Qualitative Approaches in the Study of Complex Processes

When studying social phenomena, scientists have to make a crucial decision: whether to focus on the complexity of social processes and use qualitative methods or to make broad, homogenizing assumptions about cases and document generalities—patterns that hold across many instances—which lead to quantitative strategies.[1]

The qualitative approach emphasizes in-depth analysis of a small number of cases to show how the different aspects of each case mutually interact to form the whole case, and then compares and contrasts different wholes. The quantitative strategy, in contrast, seeks broad patterns across many cases by studying a small number of dependent variables across many cases and drawing statistically valid inferences.

Qualitative research rests in part on the following assumptions: (1) populations are sets of heterogeneous cases; (2) populations are often redefined as the research advances; (3) cases are configurations of many aspects that should be understood at the level of the specific instance; and (4) causation is contextual, plural, nonlinear, and nonadditive—causes may combine in different and sometimes contradictory ways to generate the same outcomes, or the same causes may have opposite effects in similar situations. In contrast, quantitative research assumes that (1) populations are homogeneous with well-defined distribution functions; (2) populations are defined prior to the collection of data and analysis; (3) explanations are heavily variable-oriented, and the individuality of each case is not relevant (that is, idiosyncrasies are considered a deviation from the common behavior); and (4) causation is predetermined and stable, and often additive and linear, making the approach insensitive to causal complexity.

Both approaches use theory and data to build representations of reality—the difference lies in the way theories and data are used. Qualitative researchers use theories to identify important issues that must be explored in fieldwork. As the information is collected, different theories and causal links are checked against collected information on a large number of variables and interactions until a satisfactory explanation has been built. Based on these results, new theories can be developed. Quantitative researchers, however, use a priori knowledge and theories to build a concise representation of the phenomenon. With this representation, they collect large datasets of a few variables. The data are then analyzed with statistical tools to find correlations, which are seen as confirmation of the theories.

In general, there is mutual mistrust between qualitative and quantitative researchers. From a quantitative point of view, case studies are seen as untrustworthy, because they are more likely than large collections of data to provide distorted representations of broad, population-wide patterns. For qualitative researchers, quantitative research often constrains the dialogue between ideas and evidence in unproductive ways (Richters 1997), especially by imposing a simplistic causal structure on the analysis. For the qualitative researcher, confidence comes from depth; for the quantitative researcher, it comes from breadth.

Case studies do not meet the standards for valid statistical inference because, among other reasons, the latter requires random selection of the cases studied. The qualitative researcher, in contrast, deliberately selects cases because they show a particular behavior. Despite their lack of statistical validity, in-depth case studies provide the basis for constructing generalizations that hold, at least, for the cases analyzed. Often these generalizations have wider relevance. For example, although Freud's study of hysteria was based on the experience of a single patient (Freud 1963), it has been considered by psychologists as representative of a very large number of individuals.

In fact, many areas of science in which experimentation is not possible (for exam-

[1]This section relies heavily on Ragin (2000).

ple, astronomy, geology, history, and social sciences) depend on the analysis of a small number of uncontrolled cases, which plays a role similar to experiments in the construction of theories (Yin 2003). The evaluation of information in the construction of the case study, including the acceptance and rejection of specific pieces of evidence, constitutes the basis for building theories of broad application.

Quantitative analysis also has its drawbacks, because any dataset can be explained by an infinite number of theories (Miller 2000; Machamer 2002); therefore, good statistical properties (for example, high correlation coefficients) do not constitute a valid explanation, as exemplified by what is known as spurious correlation. In addition, having a relatively large random sample is no guarantee that the inferences will be valid for the whole population.

The question a social scientist should answer is: which approach is more appropriate for the research's objective and the problem under study? Quantitative approaches are useful for the study of relatively stable, simple relationships that hold for large numbers of cases; examples of such relationships are responses to conditional transfers or to educational policies. Qualitative approaches, however, are appropriate for the analysis of complex relationships that change over time or space, such as political processes or economic development.

4.3 Study Methodology

The questions this report seeks to answer are how a civil society organization managing public funds for research and extension can sustain organizational innovation over extended periods, and how it can learn and adapt to maximize its impact on the agricultural innovation system. The answers to these questions will yield lessons that will help to improve the design and implementation of research and innovation programs in developing countries.

A qualitative methodology was considered the most appropriate approach because of the process's complexity, the impossibility of controlling behaviors, and the focus on ongoing events (Yin 2003). To minimize the researcher's subjectivity, test the research design, and ensure its quality, multiple sources of evidence were used, a chronology was developed, a chain of evidence was established, and key informants reviewed the draft report to check the validity of the arguments and the accuracy of the data (Yin 2003). The theoretical propositions used to build the conceptual framework were based on the literature on complexity, innovation processes (including STI policies), organizational learning, organizational cultures, and governance. The theory was then reviewed in light of discrepancies between the theory and the evidence, or when new processes were identified. This cycle was repeated until a coherent case was built.

The study comprised a single case, in which COFUPRO was the unit of analysis, and the focus was on its learning routines, its coordinating activities in the set of relevant foundations, and its relations with external agents in the agricultural innovation system. Interactions with other actors in the broader innovation system were also briefly studied.

Most of the information was collected in 47 semi-structured interviews. The interviews were complemented with informal conversations and participation in meetings, reviews of published materials (especially from COFUPRO and the foundations), and direct observation. The different sources of data were integrated and compared to identify contradictions, and when needed, some actors were interviewed more than once to clarify inconsistent information obtained from different sources.

The interviews, which lasted between 1 and 3 hours, were conducted in the first half of 2006 and were all recorded and transcribed. These transcripts formed the basis for the database of the case study. In addition, the research was enriched with 137 interviews carried out by one of the authors in 2004 for a previous study (Ekboir 2004). In both rounds of interviews, the interviewees included current and former actors that participated in the creation and evolution of the PFs and COFUPRO, including the present and all past COFUPRO presidents and executive secretaries, federal and state public officials (including all former federal secretaries of agriculture who interacted with the PFs), managers of the National Council for Science and Technology (CONACYT) who interact with COFUPRO, INIFAP's national director, the executive secretary for the National Research and Technology Transfer System (SNITT), farmers, researchers, technical advisers, private firms, and the former research director for the International Service for National Agricultural Research (ISNAR). In the states of Michoacán, Nuevo Leon, and Puebla the interviews included the presidents, managers, treasurers, and lower level employees of the state foundations, current secretaries of rural development, INIFAP's state authorities, SAGARPA's representatives, researchers, and the presidents of the state research and technology councils. To ensure that the interviewees would speak candidly, they were promised that the individual contributions would not be identified in the final report.

Based on existing information, an initial guide for the interviews was prepared. The guide was modified as new information was collected and new questions emerged. Appendix D contains the main topics covered in the interviews.

At the time this report was being written, the research team was conducting a formal survey of foundations' presidents and managers to identify learning mechanisms developed by the individual foundations. Preliminary conclusions from the survey were used for Section 6.5.

CHAPTER 5

History of the PFs and COFUPRO

The PFs have been a sociopolitical phenomenon of great importance in Mexico, influencing the relationship between, on the one hand, state and federal political structures, and farmers on the other. They have supported the transformation of public research organizations and influenced the design and implementation of STI and agricultural policies.

The history of the foundations is an example of a complex process (see Section 2.1): none of the actors who created the foundations or who directly or indirectly influenced their development were able to foresee what the final outcome would be. Decisions that were taken at a particular moment with specific objectives in mind had unexpected and far-reaching consequences. The role played by certain individuals was crucial, but so was their ability to operate in an environment that allowed them to express their innovative capabilities.

The fact that there were 32 foundations had contradictory consequences. On the one hand, the initial system was highly atomized, with too many small foundations, leading to high operating costs and imposing a low limit on the amount that could be allocated to individual projects. Some of the foundations, aware of this problem, induced the creation of a coordinating agency (COFUPRO) to improve their negotiating position with the federal government and benefit from economies of scale by consolidating various activities. On the other hand, the great variety present in the PFs allowed the emergence of an innovative group that led the set of foundations to separate functionally from government, develop a culture of innovation, and adopt a code of conduct that placed great value on community service and a strong sense of duty.

Going beyond the initial objectives of its creators, COFUPRO coordinated collective learning among the foundations and so played an important role in their individual and collective consolidation and in the creation of an institutional culture. The interaction between the foundations and COFUPRO has been characterized by the gradual transference of power from the foundations to the coordinating agency. This transference happened voluntarily (but not easily) as some individual foundations came to realize that certain tasks could not be accomplished on an individual basis. At the same time, some foundations were reluctant to lose the independence they had enjoyed since their inception. This contradiction in the collective culture had a profound influence on the evolution of the foundations.

The combination of autonomous yet coordinated foundations together with a culture of self-criticism became an effective search mechanism (creation of variety and selection) that rapidly took the PFs to the cutting edge of knowledge regarding the financing of agricultural research and extension—conscious of the limitations of existing instruments, the foundations searched for new instruments to foster innovation. Today there are very few examples on an international level that the foundations can follow. To continue being innovative, the PFs needed to build capabilities and develop learning mechanisms to systematically study their own experiences and explore new programs and projects (see Sections 2.1.2 and 3.1).

The creation of the PFs followed an international trend in which donors and multilateral institutions promoted the use of competitive grants to induce changes in public research institutions (Echeverría and Elliot 2002; Vera-Cruz et al. 2008). The ensuing institutional innovations, however, were uniquely Mexican: foreign actors had very limited influence on them.

5.1 Mexican Agriculture in the Second Half of the Twentieth Century

The history of the Mexican agricultural sector in the second half of the twentieth century can be divided into two distinct periods: from the 1960s until the debt crisis of 1982 and from that point on. Before the crisis, most developing countries followed policies of import substitution and strict regulation of domestic markets and foreign trade. These policies caused fiscal and trade imbalances that governments tried to compensate for by borrowing in international financial markets (Staatz and Eicher 1998). By the late 1970s, many developing countries (Mexico included) were heavily indebted. In 1981, the United States tightened its monetary policy. The resulting increase in international interest rates caused simultaneous crises in several important developing countries, which, in turn, spread to international financial and goods markets. The generalized crisis showed the limits of the import-substitution strategies in an increasingly globalized economy, and the import substitution paradigm lost its appeal among policymakers and social scientists. Many developing countries liberalized the most important markets (especially trade) and reduced the size and scope of government. These policies eventually became known as the Washington Consensus (Rodrik 2006). Mexico also followed these trends.

In the first period, the Mexican agricultural policy emphasized self-sufficiency in food production, exports of a few traditional products, and support for small producers. Competitiveness or sustainability were not considered in these policies. The instruments used were investments in infrastructure (especially roads and irrigation systems), subsidies for products marketed in protected domestic markets, and the creation of public research and extension institutions. A fundamental trait of public policies was their centralized and hierarchical character.[1]

The public agricultural research institutes were created at the beginning of the 1950s to provide technical support for these policies and to train human resources for the public sector. INIFAP's main objective was to help increase the output of products of social importance (such as maize, wheat, beans, and barley) and cattle ranching.[2] Significant investments were made in a public extension system to spread the technological recommendations developed by the public research institutes. The agricultural universities, such as Chapingo and Antonio Narro, were created to educate professionals for public service (especially as extension agents), while the Colegio de Posgradua-

[1]Between 1929 and 2000, Mexico was governed by the same political party, which organized a very centralized political system, including public administration. In the 1990s, the government started a progressive transfer of power to the states and began to open the political system. In 2000 a candidate from a different party was elected president.

[2]For reasons of brevity and clarity, the name of INIFAP is also applied to the institutes that preceded it.

dos educated professors for the agricultural universities (Ekboir et al. 2003).

Organized as public offices, these institutions had little operating flexibility, weak quality controls, and focused only on the productive aspects of agriculture (such as plant breeding). Furthermore, direct government interference caused research programs to change in accordance with public policies. The universities were organized following the twentieth-century European model of higher education: teachers had no obligation to undertake research or interact with the rest of society, quality controls were weak, and universities were supposed to be autonomous institutions, isolated from external influences (Malerba 1993; Dosi, Llerena, and Sylos Labini 2006). Despite this institutional setting, a few high-quality research teams emerged in various institutions.

Influenced by the political environment prevalent in those times, the research, teaching, and extension institutions developed a hierarchical institutional culture that put a premium on obedience and discouraged creativity. The hierarchical institutional culture and the linear vision of science meant that the public research and extension institutions isolated themselves from productive agents. Setting themselves apart from the institutional culture, some researchers and extension agents established close links with farmers.

In products that were technologically mature and were traded in protected markets, researchers could guess the farmers' needs and developed some useful outputs. These were used by those farmers able to absorb external information. In other words, these institutions did have a small impact, especially among small farmers who had no access to other sources of technical information.[3]

In the 1960s, North American traders contacted farmers in northern Mexico to produce fresh vegetables for export. Driven by high profits, production rose steadily for three decades. The technology was imported, initially from the United States and subsequently from Israel and Europe (Calvin and Barrios 1999; Ekboir et al. 2003). The contribution of the Mexican public sector to this process was essentially the development of irrigation and transport infrastructure. In contrast, the public research institutions did not participate, because the emerging high-value products were not among the priorities of the federal government. Besides, the impossibility of hiring new researchers and the difficulty in retraining established professionals hindered the development of new research capabilities.

By protecting internal markets, agricultural policies reduced the need for innovation and allowed the survival of inefficient production structures. Stable policies also reduced the need for the research institutions to interact with other agents and adapt to changes in the environment in which they operated. The situation changed after the 1982 crisis and especially during the 1990s, when domestic markets (including the market for land) and international trade were deregulated. The agricultural sector had to adjust rapidly to the new rules of the game and responded by changing its productive structure. These changes enabled considerable growth in nontraditional products and, contrary to expectations, also in some traditional products, including maize, wheat, and beans (Yunez-Naude and Barceinas Paredes 2003). The expansion resulted from the adoption of more productive technologies, improvements in marketing mechanisms, and public and private support for agriculture (the latter in the form of remittances from migrants).

The new political and economic environment changed the farmers' technological and commercial needs, with competitiveness and sustainability now on the top of the agenda. Attention to these factors requires new research

[3]See Section 2.2 for an analysis of the dynamics of innovation processes along mature technological trajectories.

routines, based on a systemic perspective of productive and marketing processes, and on close cooperation between the various agents of the innovation system (see Section 2.2). For example, sanitary standards for export are very strict, so to expand foreign sales of fresh fruits and vegetables, it is necessary to improve the efficiency of sanitary campaigns. But their efficiency does not depend entirely on control measures, but also on the solution to externalities and improvements in producer profitability.[4]

The technological needs of producers of nontraditional products continued to be satisfied by the private sector, especially with imported technologies, although there were some significant but isolated efforts to develop private domestic research programs. The private sector also began marketing new technologies for traditional products, especially seeds, machinery, and agrochemicals. Public-sector research institutions continued with their traditional routines and lines of research, although, in recent years, some institutions have made considerable efforts to develop ties with other agents.

The changes in the socioeconomic environment, especially the new technological and commercial needs of farmers, exposed the inflexibility of public research and extension institutions. Even though these institutions did not have the means and the incentives to participate in the dynamic innovation processes, a few researchers (as individuals) made important contributions.

When the inability of the research institutions to adapt to the new reality became apparent, the government and the farmers began to question their very existence. Seeking to increase their efficiency and promote linkages with other private and public agents, the public research and extension systems were reformed in the 1990s. The two most important instruments used were new financing mechanisms (including the reduction of treasury grants and the introduction of competitive grants) and the enactment of laws to increase the system's flexibility, especially the science and technology laws of 1999 and 2002 and the Sustainable Rural Development law of 2001.

The aforementioned reforms were an incorrect response to a correct diagnosis of the prevailing situation of the research and extension institutions (particularly their weak contacts with farmers). The reforms failed for two reasons. First, they were based on a linear vision of science instead of a systemic view of innovation networks, and as a consequence, they did not solve the problem of insufficient interaction between researchers and other innovative agents (see Section 2.3). Second, the changes were initially implemented at the level of the regulating institutions (SAGARPA and CONACYT) and not within the research institutes. Starting in 2000, the main agricultural research institutions became Public Research Centers. This change gave them more autonomy in the management of resources, but did not bring corresponding changes in the research methods, the incentives offered to researchers, or interactions with other agents in the innovation system.

Change in the research institutions progressed slowly until 2004, when the threat of closing down these institutes and increased pressure from SAGARPA accelerated the changes. Nevertheless, the transition has been slow and progress uneven because of variability in research budgets, the autonomy of universities (which isolates them from external pressures), labor laws and budget restrictions (which hamper the recruiting of profession-

[4]An externality is generated when one producer imposes costs on another, and the latter has no legal or market mechanisms to be compensated for the damage. An example is a farmer who does not control fruit flies. His orchard becomes a focus of infestation for his or her neighbors, who cannot force the producer to control or to compensate them for their losses. Positive externalities arise when, instead of damage, a benefit occurs, for example, the presence of a bee-keeper near a cornfield.

als with new skills), unclear new institutional mandates, the inertia of a research culture that does not favor interaction with other agents of the innovation system, deterioration of human capital, and the introduction of inadequate incentive schemes.[5]

The changes in the extension system were more drastic than those in the research system. In the 1960s and 1970s, extension was considered to be an important instrument to promote economic development. However, in the 1980s and 1990s, following an international trend (see Section 2.3), public investment in extension was considerably reduced because of the apparent lack of impact, together with structural adjustment policies. The reasons for the distrust of public-sector extension lay in the significant influence SAGARPA had on the extension institutions and in the paradigm shift from self-sufficiency in food production toward agricultural competitiveness and sustainability (Ekboir et al. 2003).

Public extension structures were dismantled in the 1990s and were replaced by different mechanisms to encourage the development of a market for technical services. The results, however, were not positive because of implementation problems and the structural limitations of small firms (Ekboir 2004). The most important program to develop a market for technical services (Program for Capacity Building in Rural Areas, known as PRODESCA) was implemented as part of the Alliance for the Countryside.[6] Private firms, research or teaching institutions, and NGOs could submit to the program a project to provide technical services in a particular location. The project was required to clearly identify the types of farmers that would be served and the technologies that would be "transferred" to the farmers. Once the project was approved, the recipient institution would hire professionals to do the fieldwork. These professionals, in turn, had to contact the farmers and convinced them to join the program. Often farmers would join not because they were interested in the technologies offered, but because they expected other subsidies to be made available through the program (Ekboir 2004). PRODESCA had a top-down structure that allowed farmers little participation. The difference between PRODESCA and the public programs it replaced was that the extension agents were now hired by an intermediary instead of being public employees.

Parallel to public-sector efforts to rebuild agricultural extension, several private mechanisms of varying efficacy emerged. Many input suppliers offered their clients technical advice. Commercial farmers often contracted specialized technical services. Nevertheless, small farmers did not have access to these markets because of structural limitations that characterize this type of firm (Shapira 1999).

The experience of the Ranchers' Groups for Validation and Transfer of Technology (GGAVATT) is particularly illustrative of the emergence of valuable local experiences and the problems of scaling them up (Ekboir 2004). In the early 1980s, an INIFAP researcher specialized in cattle crossbred Zebu with European breeds in his father's field. The neighbors were interested in these new animals and soon organized a group to discuss technological matters.

Following the experience of the Argentine Regional Corsortia for Agricultural Experimentation (CREA) groups, a few INIFAP researchers induced the creation of similar groups. INIFAP

[5]In many research institutions, the incentive structure gives greater weight to publication in indexed professional journals. To publish in these journals, researchers must explore the frontiers of international knowledge, which is not necessarily what local farmers need. Thus these incentives deepen the separation of research institutions from other actors in the innovation system.

[6]The Alliance is a support program for agriculture. It is financed mainly by the federal government with small contributions from the states and the farmers. Appendix B briefly describes the program.

strongly objected to these actions, claiming that they were extension activities and thus beyond the institute's mandate. The researchers responded by organizing the meetings on Saturdays so they would not be accused of using their working hours for unauthorized activities.

Early in the 2000s, the federal authorities acknowledged the good performance of the few existing GGAVATT and mandated that the methodology be used in all publicly supported extension activities. This strategy led to an explosion in the number of GGAVATT. But most of the new groups were created by the initiative of the technicians hired to organize the groups and did not respond to the farmers' goals. The result was that only a few of the new GGAVATT improved the technological level of their members.

5.2 Creation of the PFs

Farmer support for agricultural research in Mexico has a long history. Although their financial contributions were never significant at a national level, they were considerable in some states. For example, farmers in the Yaqui Valley, in the state of Sonora, have interacted actively with Norman Borlaug's team since the 1950s.[7] Following this experience, in 1969, Sonora farmers created a civil association, Foundation for Agricultural Research and Experimentation (PIEA), to support agricultural research, mainly development of high-yielding grain seeds. To date, PIEAs are financed by voluntary contributions paid when applying for irrigation permits.

Following this example, in the following decades INIFAP pushed for the creation of similar associations to support agricultural, cattle, and forestry research in other states. In 1995, there were 56 PIEAs, though only 31 were active (Polanco Jaime 1996), but only a handful contributed significant resources to research.

The great majority of these PIEAs failed for two reasons: first, the political climate did not allow free expression from civil society organizations, and strong interference from INIFAP prevented farmers from developing any sense of ownership. This sense of alienation did not encourage farmers to contribute. Second, because INIFAP had in general little interaction with farmers, responding essentially to SAGARPA's mandates, most farmers were not satisfied with the institute's services.

There were a few successful exceptions (including the PIEAs of Sonora, Sinaloa, and Guanajuato). Three reasons can explain their success: (1) because many farmers were not poor and valued technology, they had the means and the will to buy technology, domestically or abroad; (2) commercial farmers in irrigated zones were relatively few (which facilitated collective action) but were sufficient to muster resources for an adequate contribution; and (3) the irrigation-permit system provided an efficient fee-collection structure.

An important experience previous to the PFs was the Technological Foundation of Sinaloa, created during Francisco Labastida Ochoa's tenure as governor (1984–89) and directed by Jorge Kondo López. This Foundation managed public resources and funded various technological projects. Two examples were the acquisition of nozzles to apply ammonia in Guasave and the installation of a greenhouse in Culiacán to supply hotels and gardens with ornamental plants.

In 1995 Labastida took office as federal secretary of agriculture. He appointed Kondo director of INIFAP and suggested replicating the Sinaloa experience in other states, with funding from the Alliance for the Countryside. This action had profound consequences for the later development of the PFs (see Section 5.3).

While they were considering the profile of the future PFs, Labastida and Kondo visited

[7] Norman Borlaug was a wheat breeder known as the father of the Green Revolution. In the 1960s he developed high-yielding varieties that were massively adopted in South Asia. In 1970 he received the Nobel Peace Prize for his contribution to reducing world hunger.

Fundación Chile. Carlos Arellano Sota, who at the time was working for the regional office of Food and Agriculture Organization of the United Nations (FAO), suggested that a better model was the Fundación de Innovación Agraria. This Foundation is managed by the scientific community with farmer participation. INIFAP promoted this model in Mexico, but the foundations that eventually emerged were managed by civil servants and researchers; later management shifted to farmers with participation of the scientific community. Over the years, the scientists took just an advisory role. These changes in the original model were a search and adaptation mechanism used by the first actors in the process, and they illustrate the flexibility that later characterized the development of the foundations. These mechanisms eventually became one of the foundations' greatest strengths.

As stated in Section 5.1, during the 1980s the public questioned the isolation of agricultural research institutions. The PFs were seen by the authorities as an instrument to transform research from a supply-driven system to a demand-driven one. Yet, as explained in Section 2.3.6, a demand-driven system does not solve the fundamental problem of the Mexican innovation system, which is the lack of interaction between researchers and other innovative agents.

The other three basic objectives in creating the foundations were: (1) in line with the process of political opening, allow the foundations to have a direct link with the state's farmers and to convey their research needs and priorities; (2) increase flexibility in the use of research funds, freeing them from public-sector controls; and (3) increase INIFAP's operating funds. Prior to 1996, INIFAP barely had resources to cover salaries and operating expenses.

Throughout 1996, Kondo visited every state in Mexico, negotiating the creation of the state PF with the governor and local producers. Several governors, state secretaries of rural development, and producers connected with the well-functioning PIEAs objected to the idea of a foundation, because it would compete with the PIEAs. Among them were Vicente Fox, at the time governor of Guanajuato, and his secretary of rural and agricultural development, Javier Usabiaga Arroyo. After some discussions, both endorsed the idea of the PFs and supported it vigorously. The Sonora PIEA also objected to the creation of the foundation because of its long history and because it owned assets, which it feared would be transferred to the new foundation. It was eventually decided that the PIEA would continue operating but would share part of its resources with the PFs.

The process of negotiation implicitly acknowledged that power within the PFs was distributed among federal and state agents, and, to a lesser extent, private actors. Although initially the public sector (especially the state) dominated the foundations' boards, gradually the farmers' power increased (see below). This process contrasts with the experiences of other competitive funds, both national and international, in which structures were imposed and the public sector continued to dominate them.

In February 1996 the first PF was created in Sinaloa. The process of creating the foundations continued until 1997, by which time every state, with the exception of the state of Mexico, had created a foundation. Until 2009, the public funds in the state of Mexico were administered by the State Institute for Research and Training in Agriculture, Forestry and Fishing (ICAMEX).

Once the creation of a foundation had been decided upon, the governor would invite local farmers to join the board of directors. There was no one definite criterion for choosing whom to invite, but in general the governor invited farmers who had gained some recognition as technology leaders and, occasionally, also a certain degree of social or sectoral representation. Apparently this process repeated the history of the PIEAs, where the public sector created farmer organizations. In fact, in most states the result was similar: the foundations remained controlled by the governor or INIFAP.

But there were two substantial differences with the PIEAs. First, whereas the PIEAs were financed only by the farmers, most of the PF funds were public (state and federal). Second, in some states a few farmers asked for more independence. It was greatly to the credit of federal Secretary of Agriculture Usabiaga and a few governors (and a reflection of the new social climate that was spreading through the country) that they did not abort the process when they saw that they were losing control. In most states, nonetheless, governors did resist ceding control to the farmers.

From the start, the PFs had to follow rules dictated by SAGARPA.[8] Initially, these rules indicated that at least 50 percent of resources should be allocated to INIFAP. The way in which these resources were handed over varied among states. In some cases, the funds were simply transferred, without any control over how they would be used. In other states, the PFs demanded documentation about the projects and actions that were to be implemented.

Many researchers resented having to negotiate their operational funds with the foundations, because they were accustomed to work without accountability, and they felt they were entitled to the resources. One of the arguments used to justify this attitude was that the operational rules guaranteed them the resources; therefore the PFs had no authority over them. In some cases, they considered it to be a personal offense when their projects were rejected or when they had to document the use of the resources they received. This attitude reflected a culture common to researchers educated within the linear vision of science and a top-down political system: that as creators of knowledge, they were accountable only to their peers and to SAGARPA.

5.3 Creation of COFUPRO

In the terminology of complex systems (see Section 2.1), the emergence of COFUPRO and the PFs' capacity for change were the result of the creation of sufficient variation (32 PFs) and an efficient self-organization process (personal affinities and a sense of duty) that worked as a mechanism for the selection of routines, defining what was acceptable and what was not. The fact that there were many independent but strongly linked foundations also functioned as a mechanism for the exploration and diffusion of organizational routines. Nevertheless, the effectiveness of this mechanism was reduced, because (1) information exchanges were informal; (2) no methods for systematic search were developed; (3) there were no procedures to force the individual foundations to adopt best practices; and (4) the activities of the PFs were poorly structured, so that results depended to a great degree on the personalities of those in positions of responsibility. The bases of the organizational culture of the PFs were also laid in this period; some elements of this culture still persist, albeit with significant changes.

5.3.1 Differentiation of the PFs

Influenced by internal and external factors, the PFs began a process of differentiation and self-organization from the very outset. Among the external factors, the local political climate and the organizational experience of the state's farmers were particularly relevant; the internal factors included the personalities of board members and the managers' backgrounds. In this initial period, the most innovative foundations developed a Club culture centered on the most innovative farmers (see Section 3.2).

An implicit assumption that influenced the creation of the PFs and the selection of the governance structure was that, having solved their technology problems, leading farmers were able to guide and administer research and

[8]These rules are defined unilaterally by SAGARPA; nevertheless, the process normally includes consultations with COFUPRO.

extension programs. This assumption, however, did not reflect reality; the farmers who took control of the PFs were not trained for the tasks they were called on to perform. They were all successful businessmen, but management of the foundations required training and experiences different from what they had gained in their own companies. In particular, the roles of the president and the manager had to be clearly defined, so that a professional management structure could be put in place (see Section 6.2.1) and an adequate capacity for STI policies could be developed (see Section 2.3).

To deal with this handicap, a few farmers soon began to talk informally to one another about the administrative methods they had developed. These farmers had a clear sense of duty as well as an independent attitude toward the agricultural secretaries (both state and federal) and INIFAP. The innovators devoted substantial amounts of time to the consolidation of their foundations, displacing less committed farmers. In this way, they progressively took control of the boards. The informal channels of communication enabled them to determine their shared interests and reinforced their commitments. Personal affinities and shared visions about their role were powerful incentives for these innovators to form a power bloc within the set of foundations, while the other presidents had no incentives for collective action. Through a careful management of institutional processes, they succeeded in convincing other farmers of the benefits of political independence and separation from INIFAP. They were frequently opposed by other council members.

At first, the board of directors of each PF was a diverse group consisting of farmers, a representative from INIFAP, the state rural development secretary, and SAGARPA's delegate in the state. In Michoacán in 1997, farmers in the board negotiated with the INIFAP local director to give up his seat in the board; the farmers also negotiated with the other public sector representatives that they would have a voice but no vote. It was further agreed that every research institution would be treated equally—in other words, INIFAP's preferential status was eliminated. The public-sector representatives accepted the changes to isolate the foundation from the political difficulties that were affecting the state. These changes gradually spread to all the PFs, though not without some resistance.

Several PFs set up intrastate regional councils to facilitate farmer interaction.[9] However, these councils did not function well for a variety of reasons: in some states they were closely tied to public servants; in other cases, they copied the structure of the PIEAs (in fact, in some cases, they were the same group of people); some councils were too heterogeneous, because they included farmers with differing interests (for example, cattle ranchers and avocado growers); and finally, there was no clarity as to how to identify the concerns of local farmers. In time, these councils were replaced by state product-system councils.[10]

5.3.2 Creation of COFUPRO

In 1966, Raul Ovando Rodriguez, INIFAP's regional director, linked together a group of PF presidents from central Mexico. They discussed common research needs and decided to organize multistate research projects. Ovando also recognized the need for these regional

[9]These regions were within a state and should not be confused with the regions into which COFUPRO subsequently divided the country.

[10]For several decades, the Mexican government tried to strengthen agroindustrial chains. During the administration of President Fox (2000–06), SAGARPA brought about the creation of national and state councils by products and chains. Only farmers join the councils by products, whereas the chain councils include both farmers and other actors. In general, these councils are only formal structures and have not contributed to strengthening competitiveness. The chains are called either product-systems or productive chains.

projects and supported the farmers by putting them in touch with INIFAP researchers, so that the latter could suggest solutions to the problems identified.

From this experience, the foundations from the states of Jalisco, Michoacán, Guanajuato, Queretaro, Hidalgo, Mexico, and Puebla saw the need to create an office to represent all PFs in their dealings with SAGARPA and INIFAP. A meeting of all the foundations to create this office was soon organized. The growing independence of the PFs worried INIFAP and SAGARPA, who tried to impose their own statutes on the new organization and install their candidate for president. In the founding assembly, the independent presidents successfully maneuvered to have the SAGARPA initiatives rejected; the proposed statutes were not approved, a commission to write alternative statutes was appointed, and election of the president was postponed.

Six months later a president was selected. Two candidates competed for the position: Armando Paredes Arroyo Loza (president of the Queretaro Foundation), representing the independent farmers, and Armando Carrillo Soza (president of the Sinaloa Foundation), INIFAP's candidate. Before the election, Paredes lobbied a number of foundations and secured the support of the majority. A few days before the election, Carrillo negotiated abandoning the competition for the presidency in exchange for being appointed vice president. In the assembly, the new statutes were approved, the name COFUPRO was chosen, and Paredes was confirmed as president, with Gonzalo Torres Arellano as secretary. Both factions embraced the idea—also strongly supported by INIFAP and SAGARPA—of an association independent from the federal and state governments.

From the start, COFUPRO had six objectives: (1) contribute to the institutional strengthening of the PFs; (2) create a database of available technologies; (3) foster interfoundation communication; (4) negotiate with the federal authorities; (5) raise funds; and (6) develop strategic alliances with other research-financing organizations.

Not all PFs were convinced of the need for a coordinating agency and in particular of having to contribute to its funding.[11] On various occasions some PFs failed to make their contributions on time. When COFUPRO had no money to pay salaries, Paredes covered them personally. This level of commitment to the COFUPRO project, shared by a few other foundation presidents, was a key element in the creation of an institutional culture, setting standards that subsequent COFUPRO presidents would have to meet (see Sections 6.1.1 and 6.1.2). The original plan was to locate COFUPRO's office in the PF of its president, but it immediately became clear that the coordinating agency needed to be in Mexico City, near to the federal authorities.

5.3.3 Interactions and Convergence of the PFs

When the PFs were created, there were as many administrative models as foundations. For example, each PF adopted its own statutes and criteria for choosing officials. The result was that in several states the rural development secretary was elected president of the board. COFUPRO lobbied the state governors to convince them that the board presidents should be farmers independent of, but with good access to, the political authorities.

Because COFUPRO had no authority over the individual PFs, this process meant convincing the PFs themselves, as well as the state authorities, of the benefits of this change. One of the most useful mechanisms in this process was the organization of various strategic plan-

[11] The initial agreement stipulated that each PF would contribute 1 percent of its budget to COFUPRO.

ning exercises with the participation of serving presidents and their predecessors. In this way, a common culture was created, the more independent PFs put pressure on the others, and presidents imposed by governors became allies for change.

Conscious of his lack of experience in the administration of agricultural research, Paredes sought people with experience in the topic. In one of the strategic planning workshops, he met Jesús Moncada de la Fuente and hired him as COFUPRO's executive secretary.[12] Paredes then completed the team with Raúl Ovando as technical secretary. This type of behavior was one of the pillars of COFUPRO's institutional structure: the recognition of its weaknesses and the search for specialists who could help to eliminate them.

Moncada's and Ovando's support was important, because they were respected by authorities and researchers, they had wide experience in administrating research institutions, and they brought a vision of the role of formal research in the process of agricultural development. These contributions compensated for the lack of experience of the farmers in charge of COFUPRO's management and helped to consolidate the PFs. Issues related to extension, however, did not receive sufficient attention, and even today they are one of the weakest aspects of PF development (see Sections 2.3.4 and 6.4). Paredes was supported by the most active farmers from other foundations, who shared the same vision (see Section 5.2).

A similar convergence process occurred among a few managers who helped to establish effective mechanisms for resource management and project follow up. It is no coincidence that a majority of the managers in this group belonged to foundations whose presidents had independent attitudes toward the political establishment.

As the PFs began to consolidate, the highest officials from SAGARPA and some states began to value the contributions they received from their PFs and from COFUPRO. At times, SAGARPA and COFUPRO supported one another to curb inappropriate expenditures incurred by some PFs and to force these PFs to accept the implicit rules that the majority of the foundations were defining. Another key strategic alliance was with the Association of Rural Development Secretaries, which helped to pressure those states that had not accepted the emerging organizational culture.[13]

At that time, CONACYT had begun a process of decentralizing its operations to several regions. To leverage its own funds, COFUPRO contacted CONACYT to develop joint activities. A strategic planning exercise was organized to coordinate with CONACYT's regional systems, but the collaboration was interrupted when CONACYT abandoned the regionalization strategy in 2001.

As COFUPRO's presidents gained experience, they were able to identify various weaknesses in INIFAP and started to work to bring about changes in the institute. In 2000, COFUPRO called a meeting of some researchers to prepare a diagnosis of the problems with INIFAP and the public research system. COFUPRO then gave this diagnosis to the transition team of the recently elected President Fox, which resulted in the naming of Moncada as head of INIFAP. It also contributed to the creation of SNITT, included in the new Sustainable Rural Development law. With this support, the PFs took on two roles that had not figured in the original project: they were advisers in the design and implementation of

[12]Moncada had extensive experience as a researcher, had directed an agricultural research center that preceded INIFAP, and had been an international consultant.

[13]This association is formed by the secretaries of rural development from each state.

agricultural research and technology policies and became a voice for the interests of farmers not linked to the political establishment.[14]

5.3.4 Consolidation of an Organizational Model

At first, each PF established its own way of relating to researchers and research institutions. For example, some PF presidents considered that they had the right to receive free technical advice from the researchers. In other cases, the farmers distrusted how the researchers would use the resources they received from their PFs. Because INIFAP had serious restrictions on its management of external funds, initially the foundations directly administered the resources assigned to research projects. Some foundations also tried to influence research protocols. Differences were also observed in the management of resources: some foundations had excessive administrative expenses, and some governors were trying to use PF resources to finance development projects or for their own political benefit.

As soon as they had organized themselves as an informal group, the more independent PFs and COFUPRO began to try to curb the excesses of some foundations—in particular, they tried to establish a code of conduct that the PFs should follow. The code established that the PFs should not interfere with the researchers' actions (especially research protocols), should not administer projects themselves but only finance them, should not hire researchers as salaried workers, should identify the farmers' technical demands, and should establish strategic alliances.

The above principles were not always followed. While gaining experience in the management of their resources, some PFs realized that appropriate extension mechanisms did not exist, and they started to explore different methods to facilitate farmers' access to technical information. This exploration shows that the foundations were expanding the limits of their culture and the operational rules to identify more effective mechanisms to fulfill their mandate.

The results of these efforts have, nevertheless, been limited by the lack of effective mechanisms to evaluate the experiences and by insufficient knowledge of new paradigms for the organization of research and extension. In particular, the farmers' perception was that many technologies already existed but were not used, and therefore, the problem was one of extension. This perception was based on a linear vision of science and technology. From an innovation systems perspective, various factors can cause some farmers to fail to use a technique used by leading farmers: economic restrictions (including lack of credit), riskiness of the new technology, the inappropriateness of the technique for the possibilities and needs of the nonadopters, the inability of nonadopters to purchase the requisite inputs for the techniques, or problems of scale. In sum, the lack of adoption is probably not due to problems of knowledge about the technology but to other factors that the farmers cannot control (see Section 2.3.2).

The consolidation of operational mechanisms exemplifies the importance of exploration, selection, and self-organization. The first actions aimed at improving operational mechanisms were directed at the process of request for proposals (RFP) and the management of funds. Armando Paredes contacted the Mexican Foundation for Total Quality to help with improving the efficiency of COFUPRO and individual PFs. Thanks to their advice, the

[14]During the 71 years of one-party rule, the federal government used the public programs to build its power base in the countryside. As a result, almost all farmers' associations were affiliated with the governing party. When the opposition party took office in 2000, the new administration sought to interact with civil society organizations not linked to the previous administration.

PFs improved various administrative aspects but did not develop the capacity to analyze research and extension policies and programs. This deficit did not allow the PFs to improve their control of the projects that they financed. COFUPRO also coordinated a group of innovative managers who wrote administrative manuals for the PFs.

The total quality programs were complemented by contributions from the more innovative PFs, which developed various administrative instruments. The Jalisco PF pioneered the development of a computer program for project follow up. Sonora developed an online system to manage the RFP and to serve as the basis for the accounting procedures. Nuevo Leon prepared a database to analyze information from all PFs. In 1999 Sonora, Michoacán, and Nuevo Leon carried out an analysis of productive chains. This analysis pioneered a prioritization process that was carried out 3 years later (see Section 5.4.1).

At first, about one-third of the PFs used competitive mechanisms to allocate resources; the rest handed them directly to INIFAP. The first RFPs took place in 1997 without a unified model; each foundation worked independently, without a clear idea of what they wanted to achieve. In an attempt to improve the transparency of resource allocation in 1998, Sonora and Nuevo Leon used an RFP model designed by INIFAP. Eventually, this model was adopted by all PFs. When the Michoacán PF believed that the total quality programs were not solving their problems related to research management, they contacted CIAT to learn about the log frame.[15] Other managers contributed their administrative experience to organize internal processes; in this way, they discovered that some researches had submitted projects that had already been carried out. This discovery led to the cancellation by several PFs of many of INIFAP's proposals. Kondo and SAGARPA's delegates supported the PFs when INIFAP researchers and directors complained about the cancellations.

Funds were not transferred from SAGARPA to the PFs on specific dates (see below). The PFs had problems administering resources that were not received on a fixed schedule but were used to finance activities that had to conform to agricultural cycles. Some PFs responded by creating reserves to cushion the oscillations.

COFUPRO collaborated with a number of managers to strengthen their capacity to analyze research policies. ISNAR supported this training process.[16] These activities also helped to build the institutional culture.

Paredes worked full time for COFUPRO. This set a precedent for subsequent presidents, but it also seriously restricted the type of farmers who could run for the job (see Section 6.1). The original statutes stated that COFUPRO's president had to be the president of a PF; subsequently, the statutes were modified so that COFUPRO's president could be a former president. This change was made to preserve the institutional culture and memory by opening a space for people who had once been part of the PF, and because it was recognized that the position required a full-time commitment.

The institutional culture also mandated that the foundations' presidents should not earn a salary. This condition imposed a burden heavier than many presidents were willing to bear.

[15]CIAT is one of the 15 CGIAR centers and is based in Cali, Colombia. The log frame is a method for analyzing the logical structure of projects in general and was later also used for research projects. This method is used by a number of national and international agricultural research institutions as well as by international donors. It should be mentioned, though, that this method is seldom used to manage research outside the agricultural sector. Informal consultations and a literature review showed that most private companies and many advanced research institutes, in both developed and developing countries, do not use formal priority-setting mechanisms, including the log frame.

[16]ISNAR was a CGIAR center. In 2004 it was closed, and some of its work was absorbed by IFPRI.

Some presidents began sending their managers to meetings, which complicated decisionmaking. To compel the presidents to participate, it was agreed that they could not delegate their responsibilities; this problem of excessive delegation of authority, however, has not yet been solved. Its solution demands a thorough revision of the governance mechanisms (see Section 6.2).

In 1998 the obligation to allocate a fixed share of the foundations' resources to INIFAP was eliminated from the operating rules. Despite this change, other research institutions hardly submitted any projects, mainly because they thought INIFAP still received preferential treatment, and also because they lacked the skills to prepare project proposals and submit them to external sources of funds. The PFs organized several training courses on the preparation of proposals. Despite the training, the quality of the proposals did not increase substantially.[17] Resource allocation gradually changed, to the point that today some PFs do not finance any projects from INIFAP. More commonly, the participation of INIFAP has merely declined (Muñoz 2005).

5.3.5 First Identification of Research Demands

Since their inception, the PFs faced the problem of defining their resource-allocation priorities. In 1996 and 1997, each PF defined its own priorities with no common methodology. The board members simply met, discussed the problems they observed, and decided what to prioritize.

The first formal method that the PFs used was developed by CONACYT and was based on a restriction-tree methodology and an ex ante cost–benefit analysis of potential projects. In 1997, a first attempt was made to identify research demands using this method. In addition, regional projects were identified and some were financed jointly by COFUPRO's and CONACYT's regional funds.

In 1997, COFUPRO authorities participated in a meeting in Colombia organized by Willem Janssen, ISNAR's research director.[18] There they discussed the possibility of ISNAR's supporting the PFs. The plan, however, was never implemented because of a lack of resources. Invited by the northwestern PFs, in 1999 Janssen participated in a seminar aimed at identifying research demands, where he presented a methodology developed by ISNAR. This methodology was based on accepted practices for project design and involved the organization of forums in which the various actors in agricultural chains could diagnose their most important problems. For the next 2 years, the PFs modified this methodology to incorporate the product-systems (the method they finally implemented is discussed in more detail in Section 5.4.1). The modifications were developed by the PFs themselves with no support from ISNAR, which demonstrates their learning capacity.

The PFs recognized that ISNAR's methodology was better than CONACYT's, because prioritization was based on a broader set of criteria than just expected cost–benefit, and it created a space where different actors could contribute to the definition of priorities. This realization had very important consequences, because it opened an extremely valuable communication channel between farmers and researchers, inducing a cultural change in some researchers. It also created a communication channel between producers and authorities. This channel was much appreciated by both federal and state authorities (see Sections 5.3.3 and 6.4).

[17]Good researchers were preparing good proposals before the training, and weak researchers did not improve much.

[18]Moncada became familiar with ISNAR when he participated in a team that evaluated the center.

5.4 Consolidation of COFUPRO

During the presidential transition of 2000, COFUPRO held a strategic planning exercise in which it was decided to propose an increase in the percentage of resources that the Alliance for the Countryside devoted to the PFs (from 2.8 to 7 percent of the total budget).[19] It was also agreed to increase the contribution of the PFs to COFUPRO to 3 percent of their budgets. Given the structure of the Alliance, this decision required individual negotiations with each governor. These negotiations were facilitated by the fact that the new secretary of agriculture, Javier Usabiaga, had strongly supported the PFs during his tenure as Guanajuato's secretary of rural development; he had also been a founding member of the Association of Secretaries of Agricultural Development and was highly respected among its members.

Aware of Jesús Moncada's good work at COFUPRO, Usabiaga appointed him director of INIFAP; Carlos Arellano Sota, who had worked for the FAO regional office in Santiago de Chile, was appointed COFUPRO's executive secretary.

Gonzalo Torres Arellano, then COFUPRO's secretary, took over from Armando Paredes as president in mid-2001. Gonzalo Torres had direct access to Usabiaga and shared his views regarding the role of technology in development, the factors affecting competitiveness in Mexican agriculture, and the desirability of the autonomy of the PFs. In fact, they would often discuss the situation of the PFs. José Laborde Cancino had been an important adviser to Usabiaga ever since the latter had been secretary in Guanajuato and had also been a researcher. As an adviser, Laborde had an indirect but important influence on the consolidation of the PFs and also contributed to the improvement of the RFP process.

Both President Fox and Secretary Usabiaga valued that the PFs were civil society organizations, that several PFs had demonstrated their capacity to manage financial resources and administer projects, and that they would also be able to represent politically independent farmers in the design of sectoral policy.

During Gonzalo Torres's presidency, COFUPRO's structure was consolidated and the convergence of the PFs toward a shared model was accelerated. In addition, progress was made in the creation of regional structures and the definition of appropriate actions at the local, regional, and national levels.[20] The process was facilitated by Gonzalo Torres's coordination and negotiation abilities, the access to the secretary of agriculture, and the support of the more innovative PFs. Frequently, COFUPRO allied itself with SAGARPA delegates in the states to pressure the state governors and the PFs that were not following the collective rules and values. Nevertheless, various problems persisted, because neither COFUPRO nor SAGARPA had the means to force recalcitrant PFs to change.

One of the major problems COFUPRO faced at this stage was to normalize its cash flow given Mexico's complex and bureaucratic process of tax collection and distribution to the states. Taxes in Mexico are collected between March and May. The federal treasury secretary (also known as "Hacienda") then transfers the resources to SAGARPA, which in turn transfers them to a state trust, which finally transmits the funds to individual state PFs. The funds normally arrive at the PFs between August and September, but the date varies from year to year. The three bottlenecks

[19]Guanajuato, President Fox's home state, had increased its contribution to the state's PF to 7 percent in 1977, 3 years before the other states.

[20]These regions covered several states.

in this process are the relationship of the state rural development secretary with the president of the state PF, the willingness of the latter to transfer resources to COFUPRO, and the PF's alignment with the principles of COFUPRO.

The idea that public funds could be managed by foundations independent from government control was a great novelty in Mexico. The democratization of Mexican society was a quite recent phenomenon, and many politicians objected to this change, especially in certain states. In addition, some governors refused to give up their control of the PFs, hoping to use the resources at their own discretion. Often the differing views of farmers and governors on the role of the PFs soured their relationship; it sometimes affected the governors' relationship with COFUPRO as well. In either case, the flow of resources stopped, and COFUPRO did not receive the contributions.

In other cases, however, the governors appreciated the information generated by the PFs and the communication channels they opened with farmers. Nevertheless, very few states contributed to the PFs beyond the minimum required by the rules of the Alliance for the Countryside.[21] Gonzalo Torres visited all PFs and rural development secretaries to persuade them of the need for a consolidated COFUPRO. He also established a contingency fund to cushion the cash-flow fluctuations, thus normalizing COFUPRO's income.

Another important achievement during this period was the identification of chain and state research demands (see Section 5.4.1). In 2002 and 2003 this methodology was used to identify and prioritize productive chains at state and national levels. The information generated was used for all state RFPs, the SAGARPA-CONACYT sector fund, and in the design of many agricultural policies.[22]

COFUPRO also consolidated its role in reversing the atomization of the PFs, consolidating the RFP process, defining regional projects, establishing a mechanism to review the portfolio of all state projects financed by the PFs to avoid duplications, and participating in the SAGARPA-CONACYT Sectoral Fund.

The set of 32 PFs gave states considerable representation but did not allow the financing of large projects or those of regional importance. To solve this problem, it was agreed that each PF would transfer 15 percent of its resources to COFUPRO, which would administer a fund devoted to national or regional problems. The decision to transfer resources was discussed between SAGARPA and COFUPRO and was imposed on the individual PFs. The decision to ask the individual PFs to transfer the resources, as opposed to specifying automatic allocations in the operating rules, was made by the secretary of agriculture, who claimed that it would give farmers the chance (and the responsibility) to decide. The result of this arrangement was that a number of PFs delayed their contributions. This decision was in line with COFUPRO's culture: all decisions should be consensual, even if it meant higher transaction costs.

In 2003, it was agreed that these resources should be integrated into the SAGARPA-CONACYT Sectoral Fund, of which COFUPRO became a third partner. COFUPRO's objective in joining the Sectoral Fund was to double the resources available to finance research projects of national importance. But later CONACYT

---

[21]Some states even contributed less than the required amount.

[22]In 2000, CONACYT modified the mechanism to allocate research funds with the purpose of incorporating users' demands. Separate agreements with each secretary of the federal government and with each state were signed to create competitive funds to finance research in areas of interest to the secretaries or the states. In these funds, CONACYT contributes half of the resources and the partner the other half. The topics financed are defined by the partner, and CONACYT administers the resources. The funds constituted by the secretaries are called sectoral funds, and those constituted by the states are called Mixed Funds. At present there are 28 state Mixed Funds and 1 municipal Mixed Fund. In addition, there are 14 Sectoral Funds.

could only match SAGARPA's contribution but not COFUPRO's.[23] CONACYT's failure to comply with their obligations (from the point of view of the PFs) upset some PF presidents, and the possibility of withdrawing from the Sectoral Fund was discussed; in the end, the PFs decided to stay in.

More than 1,000 proposals were submitted in the first Sectoral Fund RFP. This number exceeded CONACYT's capacity to manage the evaluation process, and COFUPRO took responsibility for reviewing the proposals, because it had more experience in evaluating large numbers of projects.

The creation of a COFUPRO-administered fund had two important consequences. On the one hand, it created a virtual division of labor between COFUPRO and the individual PFs, as the former became responsible for addressing problems of regional and national dimension. On the other hand, COFUPRO's power increased, because it was no longer only a coordinator but also the administrator of a substantial amount of resources and a privileged counterpart of CONACYT and SAGARPA in discussions about the agricultural research agenda.

At first, each PF would issue its RFP on a different date; there was no exchange of information with other PFs regarding the topics included in its RFP or which researchers had received funds. Neither was there a common methodology to select which topics would be covered in the RFPs. This atomization of the system resulted in great inefficiencies, because it allowed researchers to submit the same project to more than one PF and forced repeated evaluations of some projects. It also allowed the same topic to be financed by more than one PF. These problems were gradually solved. To avoid duplication of projects, it was agreed to issue all RFPs on the same date; once each state had evaluated its proposals, the 32 PF managers met to discuss the projects approved by each PF. In the case of duplication, they would decide which PF would finance the project.

With the consolidation of the regional projects, a trend toward specialization emerged: COFUPRO financed larger projects of regional or national impact, and the individual PFs financed local validation of research findings and extension. Despite this specialization and the convergence of various common processes (for example, administrative mechanisms and identification of research demands), each PF preserved its independence in deciding which projects it would finance. Thus the 32 PFs as a group became an exploration mechanism. COFUPRO's and the foundations' weakness was that they did not create effective mechanisms to assess their individual experiences and to share the lessons learned. Each PF had informal knowledge of the experiences of the others, but there were no procedures to systematically evaluate these experiences.

Starting in 1999, the Alliance for the Countryside was financed by a World Bank loan. The latter imposed a degree of conditionality that reduced the effectiveness of the PFs. The most important restriction was that individual grants could not exceed US$50,000. It soon became apparent that this amount was too small to finance relevant projects.[24] When the loan expired, representatives from the secretary of the treasury objected to any increase in the limit. The creation of the regional funds allowed elimination of that ceiling for projects of regional or national importance.

[23]There was a difference in the interpretation of the commitments made by each party. The constituent agreements of the Mixed Funds indicate that the partners contribute equal parts. SAGARPA and COFUPRO understood that COFUPRO was an equal partner, which increased the amount CONACYT should have contributed. CONACYT, however, understood that COFUPRO was a third—but not equal—partner; thus its participation did not increase CONACYT's commitment. In the interviews, CONACYT managers indicated their intention to increase their participation when their budget allows it.

[24]In general, in agricultural research US$50,000 barely suffices for a few experiments.

Another problem the PFs had to deal with was that resources were allocated annually, preventing the financing of multiyear projects. Researchers were forced to divide their projects into annual subprojects, which were evaluated separately. To overcome this problem, some PFs decided to give preferential treatment to subsequent submissions of subprojects. Integration into the Sectoral Fund permitted the elimination of these two restrictions for projects of national scope, although they continued to constrain state projects.

The budget law that finances the Alliance for the Countryside requires that the latter be evaluated annually. Starting with the 2001 evaluation, SAGARPA contracted FAO's regional office in Santiago de Chile, which in turn subcontracted to local consultants. This review was the first quantitative and qualitative evaluation of the research and technology transfer subprogram. FAO has carried out these evaluations to date. These evaluations helped COFUPRO identify problems that needed immediate attention, for example, lack of alignment between identified demands and projects submitted by researchers and the need to systematize all project information.

The federal and state authorities began to realize the advantages of having independent PFs, particularly after the exercise that identified research demands. The authorities especially valued the information that COFUPRO and some of the PFs provided, as well as the channels of communication with farmers they opened. But this opinion was not shared by all officials in the federal and state governments. In 2001 COFUPRO and SAGARPA started conversations on having the former support extension activities, but the project did not prosper because of the lack of interest in the SAGARPA offices in charge of the area. However, in some states, especially Michoacán, various innovative projects were developed and financed jointly by the state PF and PRODESCA, the federal program that finances the development of professional service markets.

5.4.1 Second Identification of Research Demands

One of the actions with the highest impact implemented by COFUPRO was the prioritization of agrifood chains and the identification of research demands during 2002 and 2003. The impacts resulted from (1) the opening of multiple channels of communication among farmers, researchers, and politicians; (2) the use that SAGARPA, the states, and CONACYT made of the information generated; and (3) the use these actors still make of the updates of the original information. The methodological principles for the identification were developed by ISNAR and modified by COFUPRO and a few managers, who prepared the training materials, trained the facilitators, and managed the whole exercise. Besides the information generated, an important by-product of the exercise was the training of 48 professionals in the method for identification of research priorities.

The methodology is based on the organization of forums for each chain considered important in the state. The process involves three stages. In the first stage, farmers, traders, technical consultants, and researchers meet to describe the chain and the processes by which the agricultural products reach the final consumers, and they agree on a tentative identification of problems. In the second stage, the participants are separated into three groups to discuss their vision of the problems (farmers and other commercial actors in one group, technical consultants in another, and researchers in the third). In the final stage, all participants get together again to discuss the results of the group meetings. All meetings have a moderator who helps the groups to generate adequate information. Finally, a consultant uses the conclusions of the forums plus secondary information to write a report on the chain's research demands.

The quality of the information generated depends on the careful selection of the consultant and forum participants. In addition, farmers and other private-sector actors are often reluctant to participate actively and dedicate the necessary time to adequately characterize their problems. The researchers are more accustomed to this type of activity and find it easier to participate in forums that last for more than one morning. To accommodate these differences, COFUPRO cut the farmers' forums to approximately 4 hours, whereas the researchers' forums last for 2 days.

A serious problem with this methodology is that only chains that are already important can be analyzed and prioritized. Development and competitiveness are often based on the introduction of new products and markets, but this methodology cannot analyze products and chains that do not exist yet. For this reason, it is necessary to complement the methodology with another one aimed at encouraging the emergence of new products with strong market potential.

The identification and prioritization of productive chains was complemented with the promotion by SAGARPA of state system-product councils (in which only primary producers participate) and of national and state productive-chain councils (in which representatives of all links of the chain participate). The system-product councils became a new channel of communication between PFs and farmers, and a growing number of PFs are incorporating their presidents into their boards. It is not clear, however, just how representative the system-product councils are.

Toward the end of 2004, discussions began on the need for updating the information on research demands, and the conclusion was reached that repeating the 2003 exercise would be too expensive. As an alternative, "technological innovation units" specific for each of the most important chains were created. A small number of specialists, including farmers and researchers, participate in these units with the objective of collecting information on technological and market trends for their chains and identifying research topics. However, as the definition of priorities is now based on the opinions of fewer individuals, these are less representative of the opinions of all actors in the chain. The PFs currently consider that most of these units have not been successful, but no alternatives have been explored yet.

Even though the PFs and SAGARPA classify these exercises as identification of demands, they are in fact processes in which suppliers and demanders interact to identify opportunities and needs. Without a centralized mechanism for updating demand information, several foundations developed their own methodologies, which can change as often as every year. These mechanisms, however, suffer from various drawbacks, the most obvious being the cost of the exercise and the instability it brings to research programs because of frequent changes in priorities.

5.5 The Present Stage

In November 2004, Carlos Baranzini took over as president of COFUPRO, and Raul Romo occupied the post of executive secretary, which became vacant when Carlos Arellano was appointed executive secretary of SNITT.

The present administration has continued two of the previous administration's main policies: consolidation of the structure and internal procedures, on the one hand, and on the other, strengthening COFUPRO's presence in the Mexican innovation system in general and its agricultural system in particular.

One of the most important changes in COFUPRO's operation was the creation in Mexico City of a management team composed of a few experienced PF managers. The creation of this team formalized Carlos Arellano's reliance on these managers during his term of office. Although this team strengthened COFUPRO, it weakened some of the PFs, which could not replace them easily. This difficulty reflects a dichotomy in the organization of the PFs as a whole: although each manager's contract depends exclusively on the decision of the PF's president, the manager is operationally subject to the regulations defined by COFUPRO and the whole set of managers. COFUPRO gives informal advice on hiring a manager when a president asks for it, but the final decision on hiring and firing a manager is the president's sole responsibility (see Section 6.2).

COFUPRO's management team is helping the weaker PFs to professionalize their presidents and managers, in particular training managers in topics related to the administration of the PFs. Just as in the past, the training transfers the institutional culture to the new managers and foments the emergence of an esprit de corps.

Seeking to strengthen the capacity of the PFs as a group to analyze specific themes and to improve regional and national activities, managers were grouped into eight stable groups, each one examining common problems (for example, data processing, identification of best practice, and the design of RFPs). In addition, all managers meet every 2 months to discuss operational themes. This frequent interaction also helps to create a culture of cooperation among the managers. In 2004, the PFs put into place a common information system.

Procedural rules for the presidents have also been consolidated. Several presidents used to delegate most of their work, because they could not leave their own businesses unattended; now COFUPRO is working with those presidents to get them more involved in their PFs. In addition, to reduce learning time, it is being suggested that new presidents be elected among board members, so that they are familiar with the functioning of the PF before they take office. This mechanism would also help to give continuity to the organizational culture. Until 1985, COFUPRO had only one vice president; reflecting the greater diversity of tasks COFUPRO has been undertaking, now there are three vice presidents: one for agriculture, one for fishing and livestock, and one for the research system. Finally, COFUPRO's board was strengthened by establishing clear rules for the nomination of its 19 members (see Section 6.2).

Other areas where improvements are in progress are upgrading the communication infrastructure, and an online system developed by COFUPRO to manage the project cycles (that is, the RFPs, project selection, signing of contracts, disbursement, and follow up of implementation). This system was adopted by all PFs except the one in Sinaloa, which developed its own system.

COFUPRO is also moving forward with the standardization of processes and statutes to eliminate repetitions and redundancies. This process has entailed a gradual transfer of power from the individual PFs to COFUPRO. Although this process was resisted by some PFs, the majority of them gradually came to recognize the advantages of uniformity of operational routines. In 2005, 16 PFs were refusing to adopt the common procedures; this number fell to zero in 2008.

The philosophy and the operation of state RFPs changed substantially in 2005. The 2003 exercises to prioritize agricultural chains and research topics allowed the PFs to identify problems that required immediate attention. However, the researchers generally interpreted RFPs in a lax manner, so that they could continue with their own lines of work. The result was that the PFs often did not get answers to the problems they prioritized. Additionally, the PFs believed that the competitive mechanism concentrated the resources on a few research teams and that the quality of the research was not increasing. Similar problems have been identified in other countries (see Section 2.3.7). Instead of the traditional competition, the state RFPs for 2006 identified specific projects of significance to the farmers. The PFs then invited various research institutions to explain what strengths they have to investigate one or more of the projects identified. Institutions that had not been invited were also welcome to submit their own projects on the topics included in the RFP. Based on this information, the PFs selected those institutions most capable in each area and asked them to prepare a research project. Also contemplated was the possibility of asking one or more of the prescreened institutions to make a joint proposal. These changes represent a shift from traditional competitive funding to competitive bidding of research contracts. It was expected that the new process would allow a better alignment of research with the farmers' needs. The experience, however, has not been totally satisfactory. The main lessons from the different mechanisms is that the various procedures for the RFP (1) did not improve communication between the farmers and researchers and (2) did not help to strengthen research capabilities. They failed to strengthen research capabilities because they did not finance long-term projects, investments in research infrastructure, or the hiring of young researchers.

In years past, COFUPRO has also consolidated its participation in the Sectoral Fund and decisively influenced the definition of the subjects to be included in RFPs and in other operational aspects (see Section 6.4).

Some PFs have developed a capacity to rapidly incorporate new ideas. For example, in 2003 the Michoacán PF was introduced to the concept of innovation networks—the idea that farmers innovate by interacting with peers, researchers, and other agents in stable social networks. The foundation promptly financed a project to map the innovation networks in which state lemon growers operated and to use these informal structures to facilitate the exchange of technical and managerial information. The Sonora PF used a different approach to support innovation networks: it organized meetings between some system-product farmers and researchers to discuss their research programs in depth. In this way, a more direct and long-lasting communication channel between farmers and researchers was established. When the product-systems were well organized, the exercise resulted in a strengthening of the innovation networks.

Another example of the capacity to incorporate organizational innovations is the procedures to oversee research projects. The 2004 FAO evaluation pointed out that the PFs only controlled research expenses but not the quality of research or of the final reports. The following year some foundations relaxed the administrative controls and established mechanisms to control the research results. Additionally, the PFs oversee the projects they fund not only through written reports, but they also routinely visit the experiments and talk to the famers involved.

In years past, COFUPRO has consolidated its presence in the Mexican agriculture. Approximately 4,500 people participate in assemblies and other institutional activities. COFUPRO also developed a strong relationship with the National Conference of Governors and with the state secretaries of rural development. Various rural associations with ties to political parties have applied for COFUPRO membership in recent years, but PF officials objected, because they do not want COFUPRO to be politicized. However, some of these groups do participate in the boards of a few PFs.

Even though each PF and COFUPRO have relatively modest administrative structures, they are, as a whole, an important organization with approximately 200 employees distributed in quasi-independent units (the individual PFs). Despite their size, the PFs have not become a bureaucratic organization, thanks to the same factors that have shaped their evolution and impact throughout their history: a decentralized structure and an institutional culture that values effective action (see Sections 6.1 and 6.2), as well as the influence of some individuals. This division into independent units does, however, bring with it the duplication of certain activities and consequent higher costs (for example, each PF has an independent accounting system).

CHAPTER 6

Institutional Assessment of the PFs

The PFs as a group have been a unique organization, because they have been able to redefine their values, processes, and instruments as they learned to manage public funds for research and extension. This experience has been all the more extraordinary because the PFs have been able to sustain organizational innovation over an extended period of time. Studies of private firms have found that most organizations eventually lose their creativity and seldom regain it (see Section 3.1). This chapter analyzes three factors that have been the bases of PF innovativeness: their institutional culture, the governance structure, and the learning mechanisms they created.

6.1 Institutional Culture

As explained in Section 3.2, an organizational culture is a set of basic assumptions strongly rooted in the organization's early life (Schein 1984, 1991). Several studies of organizational culture have indicated that the leaders of an organization induce the adoption of solutions to overcome threats and sources of anxiety, take advantage of opportunities, and tackle operational conditions that the organization faces at its outset. These solutions (in particular those that worked well) create a consensus about how things should be done. The repeated use of the same solutions gradually transforms the principles hidden in them into implicit assumptions about how things are; these assumptions then become accepted as safe guides to behavior until they eventually become unconscious rules of action (Schein 1991; Leonard-Barton 1995) or consolidated routines (Nelson and Winter 1982). These behavioral rules permeate the life of the organization and affect several aspects of its behavior, especially its learning modes, the way it relates to its environment, and its governance structure.

In addition to its formal structure, procedures, operational rules, and shared goals, COFUPRO has created an organizational culture that is rooted in the codes, signs, symbols, and images that make up the collective vision of the rural sector. Among these, particularly relevant are the higher status of the PF presidents, the initial idea that the PFs were dependent on the public sector, and the linear vision of science (in particular, the superiority of researchers' knowledge and the vision of technologies as pure technical processes, independent of economic and social influences). This vision of the role of science is particularly interesting, because innovative farmers did not realize that most of the technical information they used did not originate in the Mexican public research institutions.

The codes, signs, symbols, and images evolved with the PFs and were replaced by other images, particularly the ineffectiveness of public research institutions. These codes, signs, symbols, and images were the elements that shaped the essential assumptions that became the organizational culture. This culture is a fundamental element of the sense of a common identity among the PFs, inspiring commitment to the set of PFs and its members, as well as a shared vision that clarifies what is happening and what is desirable for the PFs as a whole. The sense of belonging and a shared vision were achieved through the interactive processes of communication and learning. This section focuses on the culture rather than on the codes, signs, symbols, and images, because it synthesizes the influences of the subjacent four categories.

6.1.1 Basic Assumptions of COFUPRO's Culture

The major factor behind the creation of the PFs was the perception that developed in Mexican society in the 1980s and 1990s that public agricultural research lacked relevance (see Section 5.1). A substantial transformation of the research system from a supply-driven process to one driven by demand was seen as the solution to this problem, on the assumption that successful farmers, having solved their own technological problems, had the required knowledge to supervise programs to finance research and extension. This assumption contributed to the adoption as a fundamental principle that the members of the foundations' boards should be successful farmers. In other words, it was taken as a fact that it was the farmers who possessed the really important knowledge.

Another significant component of the culture of the PFs arose from the strong interference of the state governors in the early days of their existence. In every state, the PFs began as organizations subordinate to the state's political authority. In these circumstances, the lobbying capacity and the personal friendships of PF leaders were characteristics essential to the foundations' survival and fulfillment of functions entrusted to them. Leaders become extremely important when institutions are weak, and policies can change suddenly. This principle seems to have been recognized by PF members, and it was made clear by all the interviewees, who mentioned that being a recognized farmer with good access to the governor was an essential requirement for a president.

These presidential traits are similar to those defined as a charismatic authority in a Club or Zeus culture (see Section 3.2.3). In this culture the authority of the leader comes from his or her proximity to the center of power, and the leader's selection is not based on technical capacity. This selection criterion tends to reproduce itself at all levels and permeates power relations in the organization. Thus the power of the individual depends on his or her personal relations with the highest level of management or with external authorities.

The perception of the sources of leadership changed when the PFs gained independence. Acting out of conviction and exercising their own collective power, the independent presidents brought about reforms in the governing constitutions of all PFs. The success of the proposal to create a coordinator of all PFs politically independent from, but close to, the federal government led to this political independence becoming a key value in the overall institutional culture. This value was strongly reinforced in 2000 when COFUPRO, against the prevailing political wisdom and contrary to the will of some PF presidents, decided not to support the ruling party's candidate for the presidency of the republic. The argument was that support for the official candidate would reduce the credibility of the PFs and leave them in a weak position if their candidate lost. This decision was soon proved right and confirmed the value of political independence in the culture of the PFs.

The bases of COFUPRO's organizational culture were laid down by the concerted efforts of two groups of actors: the independent presidents, especially Armando Paredes, COFUPRO's first president, and a group of innovative managers, supported by Jesús Moncada, COFUPRO's first executive secretary. They set standards for succeeding presidents and managers. For example, Paredes dedicated himself full time to the consolidation of COFUPRO. The main strategy was the creation of consensuses; with this in mind, several strategic planning exercises were organized (see Section 5.2), and the importance of negotiations among equals was firmly established. Another element of the organizational culture was the creation, by this nucleus of innovating presidents acting with SAGARPA's support, of values, rules, and routines to govern how the foundations' authorities should behave internally and when interacting with the state and federal governments.

At the same time, several managers recognized that they shared the same work ethic and began to collaborate informally to introduce novel forms of management for their foundations. Thus the managers' convergence movement coalesced in a culture that valued knowledge, technical training, efficiency, dedication to work, and a managerial mindset that emphasizes independence as well as deference to the PF president (see Section 6.2).

Currently the organizational structure of the PFs and COFUPRO corresponds to an organic model in which committees, interfunctional teams, and work groups are integrated into a structure flexible enough to adapt quickly to changing environmental demands (see Section 6.2). These characteristics typify a Task or Athena culture (see Section 3.2.3).

6.1.2 Cultural Duality: Existing Cultures and Their Carriers

COFUPRO and the PFs developed a dual culture, where a Task or Athena culture is dominated by a charismatic figure corresponding to a Club culture (see Section 3.2.3). These two cultures belong to and are transmitted by different groups in the organization.

A Task (Athena) culture is the carrier of technical knowledge. In this culture such values as horizontal work relations are cultivated, and members act out of conviction. Individual power in the group of managers is based on skill, work capacity, and the ability to argue a case. This culture was created and is transmitted by the operational team, that is, the managers of the PFs and COFUPRO's executive and technical secretaries. It is the culture that most contributes to organizational learning.

A Club (Zeus) culture has at its heart a charismatic leader. As described above, in organizations where this culture predominates, who one knows—one's closeness to influential people—is generally more important than what one knows—technical knowledge and abilities. Many organizations are born as Club cultures around the personalities of their founders but evolve into other cultures as they face the organization's emergent necessities. This seems to be the path being followed by COFUPRO and many of the PFs. The Club culture is the carrier of leadership, the organization's legitimacy, and its external positioning. It was created and is transmitted essentially by the presidents of the PFs and COFUPRO. Its contribution to organizational learning depends on the legitimacy of the leaders in the eyes of the managers and employees.

Often, members of tasks forces need and accept a leader emerging from a Club culture. Currently in the PFs the leader's authority comes from the formal structure, but for the leader to permanently authenticate his or her power in the organization, he or she must demonstrate character, ideas, initiative, and a capacity for action. If this does not happen, the leader's fundamental claim to authority (that is, proximity to political power) is severely undermined. It is very important to understand that his or her role is to lead and not to administrate. The leader is a key player for the organization's success and needs to be an evident member of the leadership group and be aware that his or her own leadership always depends on the consent of this group. This challenge is one that COFUPRO and PF presidents currently face.

6.1.3 Consolidation of Institutional Culture and Loss of Innovativeness

By the end of COFUPRO's first presidency, the set of basic assumptions at the heart of its organizational culture had been created. The next two presidencies strengthened this culture in a process of continuity.

The greatest challenges the second presidency faced were to consolidate COFUPRO's structure, ensure that the PFs converged on a shared organizational model, and convince the PFs to make their contributions to COFUPRO (see Section 5.4). The third presidency continued the process of internal structuring of the PFs and COFUPRO, emphasizing support for the less-consolidated PFs, professionalization of the PF management cadres, intensification of interactions with external actors, and consolidation of COFUPRO's role in the national and agricultural innovation systems.

The process of consolidating the decentralized structure was managed by approaching each PF individually. This dynamic was a response both to COFUPRO's lack of formal power over the PFs and to the culture of negotiation between equals developed in the first presidency. Because of this horizontal, decentralized organization, COFUPRO's president and executive secretary have to invest a great deal of time in lobbying the presidents and managers of the PFs. The inefficiency of this process is evident in various ways: (1) in the past achieving an objective sometimes depended on SAGARPA's support, (2) there is currently a need to hold multiple meetings between COFUPRO and the PFs, and (3) basic coordination problems are sometimes impossible to solve. In parallel and for different reasons, some managers concentrated on the management of their PFs and dedicated little attention to the collective action and exploration of new modalities so essential to the first interactions among managers.

Apparently the PFs as a group have achieved a stage of maturity at which their capacity for new collective actions and the creativity of the organizations are reaching a limit. This is a phenomenon that all successful organizations go through: to the extent that they consolidate their structures, their capacity to innovate weakens (Bailey and Ford 2003; Davila, Epstein, and Shelton 2006). In the case of the PFs, the loss of creativity is because the administrative processes, especially the RFP cycle and project administration, have already attained a high degree of efficiency, so there are few gains to be made in improving these activities. The other area in which the PFs have had a big impact was in the prioritization of chains, but this exercise was too costly to be repeated regularly. The most important effects of this process resulted from better positioning of the PFs in the agricultural innovation system, the organizational learning process, and the establishment of channels of communication among different actors. It will be difficult for similar actions to have such significant effects.

A few PFs still maintain a high degree of creativity in the search for new extension mechanisms and the promotion of innovation. But these are isolated efforts, and there are no mechanisms that facilitate an effective sharing of successful experiences. As was mentioned above, the institutionalized training and sharing of experiences is restricted to operative issues: there is no common space for reflection on the merits of the different experiments to foster innovation. This deficiency stems from the evolution of the PFs, which have traditionally focused on administrative issues and not on the innovative capabilities some foundations were developing. No group has as yet appeared to push the PFs to a new collective creative cycle.

The coexistence of the two cultures (the presidential and the managerial) hinders the consolidation of operational capabilities and the diffusion of best practices, because the absolute dependence of managers on the PF presidents causes instability, inhibits commitment to the PFs as a whole (especially if a PF president is not integrated into the culture of the presidents), and renders more difficult the professionalization of technical personnel.

The legitimacy of each PF is based partially on its president's interaction with public officials. Thus a PF's independence partly depends on the governor's willingness to allow its autonomy. This dependency was a disincentive in the search for other sources of legitimacy based on, for example, other administrators of science in the state or research institutions outside the agricultural sector.

6.2 Governance of COFUPRO and the PFs

The concept of governance refers to the institutional and behavioral mechanisms that define the interactions among leaders and other actors in the organization (Pérez 2003). The analysis of governance focuses on a key set of actors, usually a small hierarchical group close to the leader. Governance systems are composed of three elements: structures, processes, and strategic axes (see Section 3.3).

To progress with their own consolidation, the PFs need both formal mechanisms and a leadership that can foster a process of continuous learning that draws strength from past experiences while giving the organization the tools to cope with the dynamics of a changing environment. For the PFs, a strengthening of governance functions will bring greater organizational maturity (see Section 6.3).

The PFs and COFUPRO see themselves as farmers' representatives; however, it is a nontraditional representation in the sense that the farmers on the boards have not been elected by a constituency, but rather they have been invited by the board to join. This practice is common for many boards. At the same time, the foundations are recognized as legitimate representatives by SAGARPA, state governments, and research institutes. In other words, the legitimacy of the PFs is bestowed by the public sector and certain private actors rather than by election by farmers. The public sector's recognition is based on the perception that the PFs are in a position to contribute their independent perspectives to the design and implementation of agricultural policies. In addition, the institutional culture of the PFs presupposes that they are autonomous from the public sector and that their main motivation is the interests of the farmers they represent. As discussed below, the definition of these interests has never been explicit, which has led some stakeholders to criticize the behavior of the PFs.

6.2.1 Governance of the PFs

The creation of the PFs was intended to create a structure that would (1) represent farmers; (2) ensure an equitable distribution of resources based on a diversity of demands; (3) follow SAGARPA's rules; (4) be flexible enough to make decisions based on the social environment's demands; and (5) integrate independent but interdependent activities, seeking integration through mutual accommodation (PF and COFUPRO 2001).

Following the categories described in Section 3.3.1, the organizational structure of the foundations and COFUPRO has evolved from mutual adjustment in the early period when rules were ill defined to a more rigid structure based on standardization of work processes and rules. Today's structure follows an organic model in which committees, work groups, and interfunctional teams are integrated. Each of these arrangements is minimally structured and is flexible enough to respond to emerging demands. Nonetheless, with nearly 200 employees, the set of PFs makes up an organization of considerable size. This design has great advantages, but it requires that those in charge of the various governance functions have skills in negotiation, harmonization, strategic thinking, leadership, and recognition from the community.

Figure 6.1 Example of a network centered on a Produce Foundation

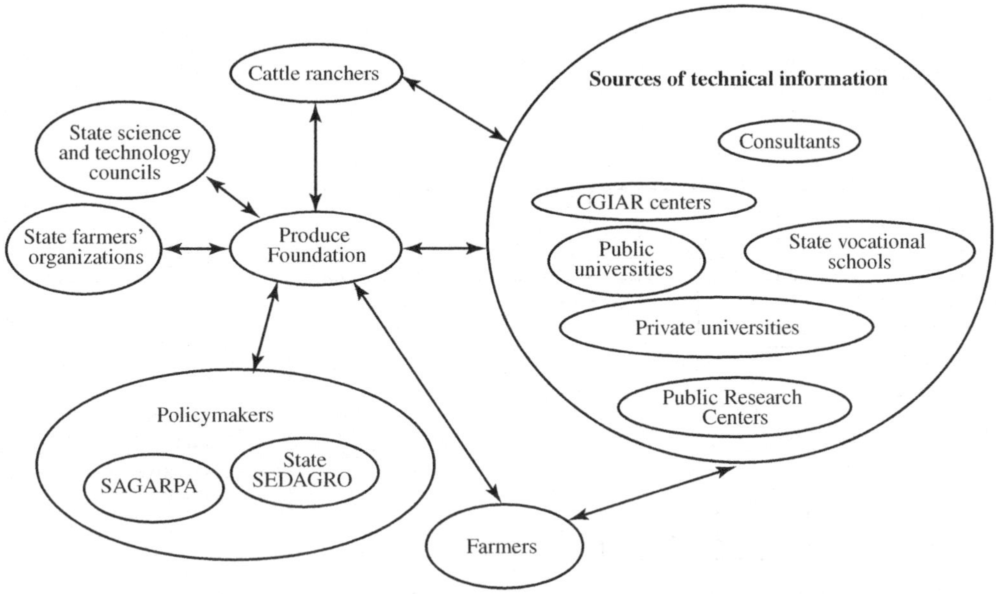

Source: Based on interviews and documentation from COFUPRO (2001).
Note: CGIAR, Consultative Group on International Agricultural Research; SAGARPA, Secretary of Agriculture, Livestock, Rural Development, Fisheries, and Food; SEDAGRO, Secretary of Rural Development and Agriculture.

Another characteristic of the foundations' and COFUPRO's present organizational structure is that they have taken a leading role in networks that integrate various agents from the innovation system, including public servants, research institutions, and private actors (Figure 6.1).

Figure 6.2 represents the current organizational structure of a PF. Each PF's government functions are the responsibility of the assembly of associates, which is made up of farmers and recognized agribusiness people. The assembly elects its new members and designates the board. There is no uniformity among the PFs regarding the criteria for deciding who is to be invited to join the board. The boards' first members did not represent any organization in particular. Although farmers participated on behalf of themselves, the civil servants represented their organizations. At present the PFs are inviting the presidents of the system-product state councils to participate on the boards; in some foundations, representatives of rural unions also participate. Although these practices raise the diversity of farmers on the board, it is not clear to what extent the state councils are representative of farmers' interests.

Since the creation of the PFs, the statutes have stipulated that government functions should be honorary (that is, unpaid), whereas operational jobs were to be remunerated. That government functions are purely voluntary means that they typically attract only businesspeople with significant resources and a relatively small time commitment to their own enterprises (such as retired individuals). With the inclusion of representatives of other farmer organizations on the boards, however, members increasingly have to devote time to their firms and all the other organizations they participate in. This leaves them with relatively little time for the PF boards and makes it difficult to uphold the cultural ideal of PF presidents and board members being dedicated to

Figure 6.2 Organizational structure of a Produce Foundation

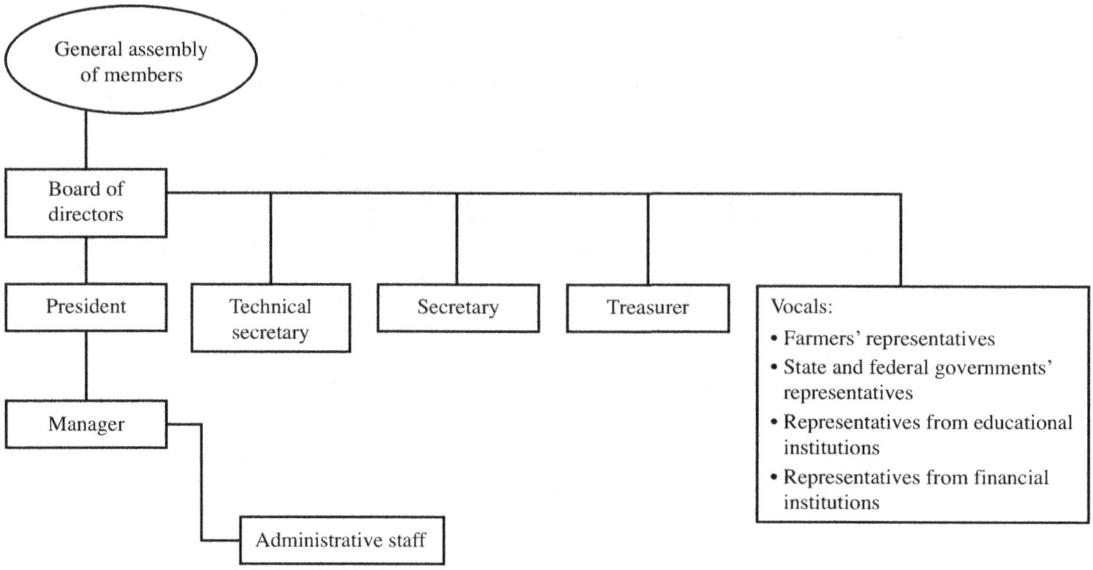

Source: Consejo Estatal de Ciencia y Tecnología de Puebla and Fundación Produce Puebla (2004).

their foundations and of the division of labor between president and manager.

That the presidents are always wealthy farmers does not imply that they have an elitist perspective: the fieldwork for this study made it clear that many presidents have a strong sense of duty and commitment to agriculture and poverty alleviation. In other cases, though, the president failed to dedicate to the PF the time it needed and delegated running the organization to the manager. The problem with delegation is that the structure and culture of the PFs indicate that the responsibility lies with the president: it is he or she who is responsible for the PF's assets, and the president is expected to exercise authority. A PF manager does not have the same hierarchical rank as the president, and a communication gap is created with other foundations and with the state and federal representatives when the manager is delegated too much responsibility.

The presidential term of office was established as 2 or 3 years with the possibility of one re-election. This rule is common to nearly all PFs. In general, boards meet at least six times per year.

The diversity of foundations resulted in different types of power structures, because there are various formulas for negotiation and cooperation between farmers and the state public sector. Despite the diversity, there was never a balance of power between the two big blocks of actors in the foundation boards: the public-sector representatives (INIFAP, SAGARPA, and the state government) and the farmers. Formally the role of most public-sector representatives is to see that operational rules are not violated. Transfer of power took place in all PFs but at differing paces and with varying intensities that reflect the capacity of some PFs and COFUPRO to generate policy proposals and administer public funds. The transition entailed changes in the sources of power (see Section 3.3.2). Initially the power of public representatives originated in their positions (legitimate power), whereas that of the innovative farmers was based on the identification of useful traits (referential power) and sometimes skills (expert power). Among the managers, power originated in skills; over the years, though, it mutated partially to legitimate power (defined by their position in the PFs as a whole). Skills are still recognized as an important trait.

The relationship between governors and those presidents who have little power can be characterized as Club culture, in which authority is legitimized by the leader's personal traits and by the devotion and emotions awakened among his or her followers. In some states, loyalty is also linked to the command of public-sector resources, especially the capacity for distributing favors or creating problems. In these structures, the administrative apparatus is made up of the most loyal disciples and subordinates, and choices are made based on the leader's confidence in these subordinates.

The PFs that achieved more independence from the public sector developed faster, fulfill their missions more effectively, and more assiduously explore new ways to promote innovation. These characteristics are due to a greater capacity for institutional innovation, based on the presence of more innovative individuals, a greater commitment by the farmers to the board, and more professional management. In those PFs that did not become politically independent, resource allocation is mainly dictated by the state governor or INIFAP's state director.

The more developed PFs share the following characteristics: (1) an administration that is autonomous yet close to the public sector; (2) organized, active farmers who insist that their demands be met; (3) a socially responsible president committed to the foundation's mission; (4) a president who thinks strategically and is eager to explore; (5) a president elected by the farmers; and (6) a professional management committed to improving operational procedures.

There is, in general, little clarity regarding some functions of the president and the manager. Because they are the highest executive authorities, some presidents have managed their PFs with the same logic they used to run their own companies. For example, they have taken upon themselves the definition of the strategic lines of work, allowing the board of directors little participation, or they have appointed managers on the basis of loyalty rather than professional capacity. Sometimes these practices have led to abrupt changes in the operation of the PF after a change of authorities.

The interviewees for this study generally agreed that the president need not know about STI policy. The presidents' and managers' lack of training in these areas, together with the fact that most managers are specialized in operational topics, weakens the capacity of the PFs to explore different means of fostering innovation. But this area is precisely where the PFs could make their greatest contributions. Traditionally the capacity for innovation in these areas was concentrated in COFUPRO and a small number of PFs. The independence of the individual foundations complicates the diffusion of best practices (see Section 6.3).

In addition, some presidents confuse the role of the PF manager with that of a private company. A PF manager is supposed to be responsible for the fulfillment of technical agreements with other foundations (generally overseen by COFUPRO). The manager is also supposed to be well versed in the field of agriculture and able to communicate with researchers and the administrators of research institutions. As mentioned before, though, in some cases managers are selected not for their technical ability but for their loyalty to the president. It is not unknown for incoming presidents to replace managers so as to be sure of their loyalty, even though the replaced manager may be professionally competent.

The managers' strong dependence on their presidents puts them in a weak position vis-à-vis the bureaucratic hierarchy of the public sector. It is highly unlikely that a manager would be able to directly call a rural development secretary or the state's SAGARPA delegate. That role belongs to the president, even when the business at hand is technical and falls within the manager's area of expertise.

The lack of definition of the manager's function in terms of technical skills has led to a situation whereby in some PFs the president is directly involved in operations, but in others the running of the foundation has been left entirely to the manager. The almost exclusive dependence of managers on presidents is a hindrance to the diffusion of agreed practices, the exploration of new methods of interaction with researchers, and the trial of new administrative routines.

The functions of presidents and managers are in general highly unstructured. This is partly due to the capacity for self-examination (see Section 5.3) that has led some foundations to explore new methods for fomenting innovation and partly to the strongly president-centered culture (see Section 6.1). It is difficult to structure tasks that are not well understood and that change rapidly or essentially depend on loyalty rather than technical skill. The literature on personnel rotation has shown that changes affect company performance when (1) outgoing staff are exceptionally effective and (2) the organizations or positions they leave have weak structures or restrictions (Coriat 2000). Some PFs fall into this category, as was demonstrated by the difficulty in replacing some managers. At the same time, a tighter structuring of the PFs could reduce their capacity for innovation, given that it requires more adherence to proven routines and less exploration of alternatives (see Section 2.1). The challenge faced by each PF and by COFUPRO is how to strengthen structures while leaving space for creativity.

6.2.2 COFUPRO's Structure

Like the foundations, COFUPRO has established a network of formal and informal, direct and indirect relations with a wide spectrum of institutions and actors (Figure 6.3). COFUPRO's highest body is the assembly, made up of the presidents of the 32 PFs and representatives of SAGARPA, CONACYT, and INIFAP; the board of directors is formed by 10 presidents who are chosen by the assembly and proposed by COFUPRO's president.

The government function lies with the president, the vice presidents, and the executive secretary. Around these positions the farmer, academic, research, and SAGARPA committees are positioned. The president is expected to play various roles: know what is going on in the foundations, take charge of political relations, legitimize COFUPRO's position, and project his or her expert and referential opinion. Nevertheless, it should be noted that the farmers are not always adequately trained to fulfill all these functions (see Sections 5.5 and 6.1).

COFUPRO's presidents are elected by the presidents of the PFs. As often happens in this type of association, there is intense lobbying before the elections. COFUPRO presidents hold office for 3 years and can be re-elected once. The presidential transition lasts for 5 months.

COFUPRO's presidential candidates must be presidents or ex-presidents of a PF. This requirement is a way of training the new president and guaranteeing continuity of the organizational culture. The implicit requirements are that a president should have a sense of duty, not be seeking personal gain, be a successful farmer, and be well connected with the political establishment (in other words, the candidate should have referential power; see Section 3.3.2). These last two conditions confirm and legitimize the president's prestige.

As for the president, the institutional culture indicates that the executive secretary should be respected by the PF presidents and managers, academics, and SAGARPA for being a good executive and should support the development of research, extension, and innovation programs and policy proposals.

Figure 6.3 The COFUPRO network

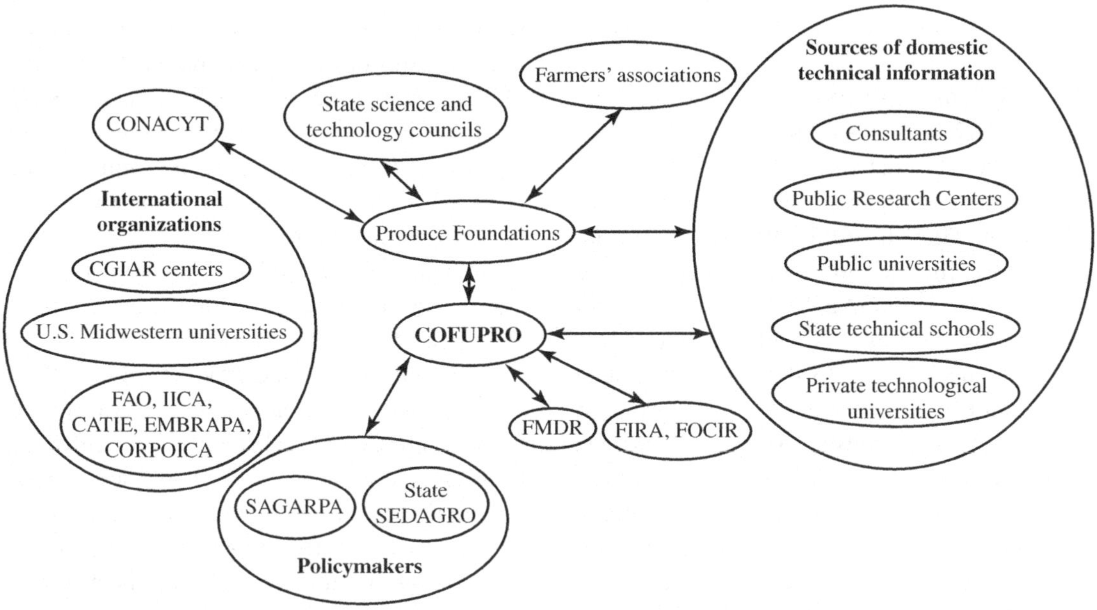

Source: Based on interviews and COFUPRO (2001).

Note: CATIE, Tropical Agricultural Research and Higher Education Center; CGIAR, Consultative Group on International Agricultural Research; COFUPRO, National Coordinator for the Produce Foundations; CONACYT, National Council for Science and Technology; CORPOICA, Colombian Corporation for Agricultural Research; EMBRAPA, Brazilian Agricultural Research Corporation; FAO, Food and Agriculture Organization of the United Nations; FIRA, Agriculture-Related Trusts of the Central Bank of Mexico; FMDR, Mexican Foundation for Rural Development; FOCIR, Trust for the Capitalization and Investment of the Rural Sector; IICA, Inter-American Institute for Cooperation on Agriculture; SAGARPA, Secretary of Agriculture, Livestock, Rural Development, Fisheries, and Food; SEDAGRO, Secretary of Rural Development and Agriculture.

Although in the past the scientific paradigm was provided by a single individual, today a solid, stable team is needed. COFUPRO and some of the PFs have recognized the limitations of the linear view of science (see Section 2.3) and are exploring new mechanisms to promote innovation in agriculture. But even at the international level there are still no universally accepted programs to find such mechanisms. Furthermore, given the complexities of innovation processes, recipes cannot just be mechanically imported without adapting them to the specific conditions of the target group. The development and follow up of these options can only be undertaken by trained professionals who coordinate their exploration with the PFs.

In the present circumstances, the most important requirements for COFUPRO's executive secretary are to be a good negotiator and be able to help the managers arrive at a consensus. Moreover, his or her good character and reputation as an expert bring recognition from the PFs as a whole. In this situation, it is important to explicitly discuss the job requirements for the position of executive secretary, especially the notion that he or she should lead the collective thinking toward a new view of science and how to interact with decisionmakers (especially in

SAGARPA and the states). It is also necessary to discuss the distribution of power between COFUPRO and the PFs, as discussed below.

6.2.3 Relationship between COFUPRO and the PFs

The relationship between COFUPRO and the PFs is one of both permanent struggle and collaboration. Though COFUPRO has no formal power over the individual foundations, in the past access to SAGARPA gave COFUPRO considerable leverage. If a PF did not follow the rules, the SAGARPA delegate in the state could halt the disbursement of funds or talk to the state governor. This informal mechanism is not efficient, because it depends on individuals rather than institutions, it can disappear at any moment, and it does not contribute to the creation of a sense of belonging to the set of PFs.

At the same time, the decentralized structure has two important advantages: (1) the PFs have a significant local presence that lets them be an effective channel of communication between farmers and the states and (2) it is a decentralized structure for exploration. The problems are (1) heavy dependence on the governor; (2) lack of an efficient mechanism to systematize and learn from the experiences of the PFs, which lowers the value of exploration; (3) a system that is too atomized and therefore inefficient; and (4) high operating and transaction costs.

With few exceptions, the qualifications of the managers and presidents of the individual PFs and the resources they command do not enable each of them to develop a fruitful vision of the role of research in supporting innovation and to effectively explore new instruments to foster innovation. This task should be the responsibility of a central organization, with specialized professionals and good connections with the individual PFs and with foreign and domestic research organizations.

The managers of the 32 PFs have organized themselves into eight stable work groups coordinated by COFUPRO. Each team is made up of four managers and prepares proposals on specific topics related to COFUPRO's various work areas. The teams meet prior to the national managers' meeting to prepare their proposals; the team leader is responsible for presenting these proposals for discussion and garnering a consensus so they can be implemented at all PFs. National managers' meetings take place every 2 months.

When a PF explored new research or extension programs, operating rules were at times somewhat restrictive. In such cases, as long as the PF could justify the program, SAGARPA was flexible regarding unfulfilled rules. But these mechanisms worked only on an informal basis and were lost in 2006 when a new federal administration took office. It would be convenient to explicitly incorporate the possible exploration of new operational mechanisms into the operating rules.

6.3 Characterizing the Learning Mechanisms of the PFs and COFUPRO

Learning is a process of repetition and experimentation that enables better and faster execution of tasks as well as the identification of new opportunities. An organization's learning capacity is a function of both individual factors (the creativity of individuals in the organization, for instance) and collective factors (the capacity to change and the organizational culture). Organizational capabilities must be built up gradually: they cannot be easily bought or copied (see Section 3.1).

The PFs as a whole and COFUPRO each have specific but interdependent spheres of action. The PFs were created to operate in the states, whereas COFUPRO is charged with coordinating the actions of the set of PFs and operating at the national level. The interdependence creates a close relationship in terms of learning: COFUPRO's learning depends not only on its own activities but also on what the individual PFs learn. The following analysis concentrates principally on COFUPRO's sphere of action.

6.3.1 Objectives, Sources, and Localization of Learning

COFUPRO is an organization oriented toward facilitating the foundations' learning, supporting SAGARPA and state governments with original data, and facilitating interactions among various actors in the agricultural sector with the objective of strengthening their innovative capabilities. COFUPRO's spheres of activity are the analysis of scientific and technology policy together with the linkage between farmers and research institutions at the national level. Given this profile, there are four major learning objectives:

- identifying the information needs of producers;
- identifying existing research capacity and researcher behaviors;
- identifying how relations among sector agents (the academic community, including institutions, researchers, and research groups; farmers; and government) influence the learning capacities of actors; and
- improving and homogenizing organizational aspects of the PFs as a group, including operations, relations between the PFs and COFUPRO, and decisionmaking.

These objectives are dynamic: they are continually adapted for various reasons, particularly the actions themselves of COFUPRO and the PFs. In this sense, to the extent that COFUPRO and the PFs learn about and act on a problem, they change the nature of the problems they will face next time. To keep this process active, COFUPRO and the PFs not only have to learn in one particular moment but they also have to develop their dynamic learning capacity.

Responding to these objectives, COFUPRO has combined internal and external learning and knowledge sources. The main internal sources were the operational and training experiences of the PFs themselves. Several external sources stand out: ISNAR, Chile's Fundación de Innovación Agraria, FAO's evaluations, and the experiences of other domestic and foreign organizations. At first, the external sources contributed important elements to the collective learning, but as the processes became institutionalized, the internal sources became more relevant. Training of the presidents and managers was particularly important. The process was essentially Mexican, in the sense that the PFs learned how to identify their knowledge needs and how to look for and adapt external information. Foreign agents (for example, multilateral organizations or international research centers) only influenced the process indirectly by providing important pieces of information.

COFUPRO's main learning activities are

- identifying the research needs of farmers and agrifood chains,
- organizing RFPs,
- administering projects,
- interacting with foundations, and
- interacting with other agents (academia, government, farmers) in the agricultural innovation system.

The relative importance of these activities changed over time. At first, the most important activities were the RFPs and project administration. Between 2000 and 2003, identification of research needs was crucial, but its demands on human and financial resources were considerable. Today's key activities are the interactions among foundations and with external agents.

6.3.2 Learning Mechanisms

Although six learning mechanisms were identified in the fieldwork, two stand out as having been crucial: identification of farmer needs and the RFPs. But there is a key weakness that also stands out: the PFs document few of their own processes and activities, especially those not concerned with formal meetings.

Identification of demands. Because in this report COFUPRO's terminology has been adopted, these activities are called demand-identification methodologies. But, in fact, they were mechanisms to link different actors in the innovation system. From the first meetings to the present day, four different methods have been used.

The first attempts to identify demands in 1996–97 were very informal, which is why they do not constitute a method as such (see Section 5.3.5). The members of various boards talked among themselves and identified problems they considered to be relevant. Between 1997 and 2001, by means of the regional research systems, an alliance was established with CONACYT. The CONACYT methodology was based on a cost–benefit analysis of the different demands identified. The most important lesson learned was the inadequacy of this second method.

Most of the lessons learned came from the third method, which was an adaptation of a methodology developed by ISNAR and improved in an exercise implemented in 2002 and 2003 (see Section 5.4.1). This method is essentially based on the organization of forums with different actors from the innovation system. In other words, the technology "supply" and "demand" are put together to identify that research with the highest possible potential. The exercise resulted in the identification of the most important chains at the state and national levels as well as the definition of research priorities.

This mechanism required a great effort in terms of time and money, so it was decided not to use it to update the priorities. Instead, the fourth method was implemented: innovation units were established for the main chains. A small number of recognized actors from each chain participate in these units to monitor worldwide scientific and technological developments for the purpose of identifying emergent problems. The latest assessment is that these units have failed, because they have not developed close links with agents in the chains; in other words, they are traditional expert opinions.

These different demand-identification methods used over the years reflect a change in the vision of science from a supply-driven model to one that is demand-driven. Nonetheless, this separation between supply and demand still reflects a linear vision of science and does not solve the essential problem of the lack of interaction between researchers and other actors in the innovation system (see Section 2.3).

Beyond the identified priorities themselves, most of the lessons learned through the exercise were only identified by an evaluation of the PFs conducted in 2004 (Ekboir 2004). In other words, most of the lessons were tacit knowledge that the PFs were not aware they had. An obvious lesson from the 2004 evaluation is that the 2002–03 exercise was useful but too expensive. This result leaves the PFs without a clear methodology for defining priorities and without the capabilities to develop a new one. Another lesson identified by the consultancy was that the exercise was more important for opening channels of communication among agents in the innovation system than for the priorities themselves. The reason is that the priorities have been ephemeral and difficult to update, whereas the communication channels have induced important and lasting behavioral changes in several agents, the PFs included (see Section 6.4).

Evolution of the RFP. An RFP is an open invitation issued by a funding agency to some actors to submit project proposals. The RFPs evolved as COFUPRO and the PFs became more independent of public research institutions and attained a better understanding of research and extension processes. During the first 2 years of their existence, barely one-third of the PFs used RFPs to allocate funds. The topics were generally very broad and therefore allowed researchers to continue with their own lines of research, regardless of farmers' needs. After the demand-identification exercise, RFP themes became more focused. Nonetheless the PFs continued to find that many researchers were unwilling to address the farmers' needs and tried to continue with their traditional research lines.

In 1997 and 1998, with the support of CONACYT's regional funds, the PFs of Nayarit, Jalisco, Michoacán, and Colima began to explore the possibility of undertaking joint projects. This process was repeated in Sonora and four other northern states. These exercises led to the implementation of regional projects. In 2000 these were extended to all the states. This activity was a key forerunner for the creation of the regional projects coordinated by COFUPRO (see Section 5.5).

During this period much effort was dedicated to analyzing the pertinence of the proposals, but the analysis of scientific quality was weak. There were also problems with the analysis of extension projects and project control, because only spending (and not research quality) was controlled directly by the PFs (Ekboir 2004). Starting in 2005, some PFs began to accept the research institution's control over expenditures and focused their efforts on quality control of the final reports. Although this stress on the quality of results is an important improvement, there is still a need to incorporate incentives for researchers to interact more productively with farmers.

As the regional projects were consolidated, the idea that it was necessary to institutionalize a division of labor between the PFs and COFUPRO started to gain strength. It was therefore decided that COFUPRO would finance projects of regional or national importance, while the PFs would restrict themselves to funding local validation and extension projects.

Over the years an RFP cycle has been developed, which is documented in the procedure manuals of the PFs: (1) identification of research and extension needs; (2) fine tuning of terms of reference; (3) diffusion of the RFPs; (4) analysis of concept notes; (5) requests for complete project proposals; (6) analysis by the technical committee; (7) requests for technical and financial revisions; (8) communication of results; (9) signing of agreement; (10) transfer of resources; and (11) project follow up (annual in situ visits and bi-annual reports).

Once the deficiencies of the RFPs were recognized, an important change was introduced in 2006: the RFPs were issued only for clearly defined projects, emphasizing controls on execution times and expected results (see Section 5.5). Emphasis was also placed on supporting interinstitutional projects.

The lessons learned from administering these RFPs were used in the SAGARPA-CONACYT Sectoral Fund (see Section 5.4). Although the participation of COFUPRO in the work group (made up of COFUPRO, CONACYT, SAGARPA, and SNITT) and in the Sectoral Fund committee has enabled an improvement in the fund's work, it is still organized on the basis of a mechanistic view of the research, extension, and innovation processes.

Despite the great effort to identify research needs, the FAO evaluations found little congruence between needs and projects funded. The recommendation was that efforts should be made to eliminate this discrepancy (Muñoz 2005). Complexity theories offer an alternative explanation of this divergence, suggesting that the discrepancy could be due to the difficulties of correctly identifying the main and emerging trends that drive the dynamics of the process (see the subsection "Defining Strategies in Complex Systems" in Section 2.1.2). Instead, then, of forcing researchers to follow the priorities, the discrepancies can be viewed as a way of exploring possibilities and thus making the RFPs more flexible.

The administration of the RFPs has now reached a high level of maturity, and there is no need to continue investing significant amounts of resources in its improvement. Instead, it is necessary to strengthen the capacity for the exploration of new alternatives of interaction with research institutions and to define new mechanisms for extension and support for innovation.

Although the PFs have made a considerable effort to improve the execution of research projects, less progress has been made with extension projects. It is generally agreed that traditional extension methods are not very effective (see Section 2.3.4), but the PFs have not found mechanisms to improve the effectiveness of these projects. However, some PFs have, on an individual basis, experimented with new extension methods. It is necessary to create mechanisms to facilitate the interchange of experiences and training in new extension approaches.

Meetings and visits among PFs. COFUPRO and the PFs have developed mechanisms to identify and socialize best practices. Examples are managers' meetings, visits between foundations, and coordination of state and regional projects. Throughout the existence of the PFs, meetings have been a major learning mechanism. In the early years, planning and mission-identification exercises included both new PF presidents and their predecessors. These were a potent way of socializing knowledge and developing a common work methodology (see Section 5.3). At the same time, through meetings and training sessions organized by COFUPRO, some managers with similar interests got in contact. This group played an important part in the institutionalization process observed in recent years. Periodic national meetings were also essential to the process of learning about the structuring of RFPs and socializing that knowledge.

Another mechanism that has enabled the socialization of knowledge is the participation of ex-presidents on the boards. This participation enriches the meetings and helps to perpetuate knowledge acquired by previous directors.

From the beginning, exchange visits have been a mechanism for learning, socialization of practice, and generation of a common culture (see Section 5.3.4). There are two problems with this practice: (1) the interchange of experiences is restricted to only those foundations that are willing to learn and (2) it is not a systematic mechanism for the exploration of best practices.

Training. The first training exercises involved the presidents, but in recent years training has been concentrated almost exclusively on the PF managers and on operational matters. This emphasis has left a significant vacuum in the capabilities of the PFs, because there has been no development of a systemic capacity to analyze methods to promote agricultural innovation.[1]

COFUPRO has strongly emphasized the training of managers with various different courses. It is not clear, however, to what extent there has been adequate definition of what type of training is needed by the management team. Some interviewees indicated that there was less need for high-level general management courses than for a more specialized training tailored to the specifics of research, extension, and the promotion of agricultural innovation, including the development of the ability to interact with researchers and farmers. Moreover, managers have not been trained in recent advances in research, extension, and innovation management.

The use of training as an instrument of change was also apparent in interactions with researchers. Various courses were offered to them, including several on research proposal preparation and the log frame. The log frame is

[1]A few exceptional PF presidents and managers have developed some incipient capabilities in this subject.

a methodology to prepare projects based on a clear definition of ends, means, results, and conditions to fulfill for project completion, as well as potential problems (see footnote 15 on page 67). To date, some 2,000 researchers have participated in these courses. The log frame facilitates project evaluation, identification of potential products, and management improvements. Although this method has been developed for clearly defined projects, some research institutions and donors have used it for research projects. The log frame has not been formally evaluated, but the literature on innovation systems indicates that it is too structured to foster more than minimally incremental innovation (Davila, Epstein, and Shelton 2006). Beyond the advantages for the PFs of a standardized presentation, however, it is not clear what benefit the log frame brings to science administration, because it reflects a mechanistic view of the research process (Ekboir 2003). Moreover, no systematic efforts were made to enable researchers to incorporate new mechanisms for interaction with other actors in the innovation system.

Hiring or interacting with key individuals. From the start, hiring or interacting with key individuals has been one of the most important learning mechanisms for the PFs, contributing to the development of the capacity to identify and assimilate new ideas and practices. A case that illustrates this learning mechanism was the hiring of Jesús Moncada as COFUPRO's first executive secretary. He brought with him knowledge of the agricultural sector as well as domestic and international contacts. This practice has been repeated in several PFs that have appointed ex-researchers or INIFAP managers to managerial posts. These professionals brought knowledge not only of the sector and its agents, but also of practices that have succeeded or failed in other organizations. Interaction with international consultants, as in the case of William Janssen, helped in learning international best practices.

It should be stressed, however, that the utility of interacting with outside experts depends on the internal capacity to assimilate their knowledge (see Section 3.1). This capacity has to be built gradually into the structures to prevent its loss with personnel changes, which is why training managers (who generally outstay their presidents) and constructing spaces for collective thought are so important.

Experiences of other organizations and informal networks. Since their inception, the PFs have sought interactions with domestic and foreign organizations able to contribute their experiences. Kondo and Labastida made their first international visit to the Chilean Fundación de Innovación Agraria (see Section 5.2). Visits to similar organizations in other countries followed.

Some of the farmers in the PFs are active members of other associations, which expose them to other organizational models. This exposure has helped the PFs incorporate new practices: for example, Gonzalo Torres participated in the pig farmers union, where each member paid a fixed per capita quota. This experience helped him convince the PFs to contribute to COFUPRO. Marta Xochitl Flores Estrada contributed her experience as manager of an *ejido;* Armando Paredes, his business experience; and Carlos Baranzini had as much experience with business as he had with farm management.[2]

The PF managers also contributed the experience they gained in other organizations. For example, Francisco Maldonado, former Michoacán manager, had banking experience,

[2]The *ejidos* are a kind of collective farm where the land is collectively owned by all *ejido* members, but each member manages his or her own plot independently. Until 1992, *ejidos* could not sell or rent their land. A constitutional amendment in that year allowed for the privatization of the *ejidos*.

which helped him diagnose the lack of clear procedures and prescribe remedies. Mauricio Mora from Puebla had research experience. Raúl Romo contributed his knowledge of sectoral policy in a state where agriculture was booming.[3] These examples clearly show that an organization's learning capacity depends on its people and an environment that allows experimentation and change. Typically when policies and programs are designed, great efforts are devoted to analyze the organizational structures, but it is usually overlooked that without capable people and incentives for risk taking, failure is almost certain.

Another source of knowledge about similar organizations has been the participation in the SAGARPA-CONACYT Sectoral Fund. This participation has had various effects: (1) by allowing access to the experiences of other Sectoral Funds, external learning was combined with in-house practical experience; (2) COFUPRO knowledge was transferred to other sectors; and (3) COFUPRO raised its profile outside the agricultural sector (see Section 5.5). COFUPRO is a member of the fund's workgroup and committee (the fund's highest administrative organs) and is responsible for framing the terms of reference for the RFPs. To do this, it uses its data on research demands.

COFUPRO has participated actively in the Sectoral Fund, a fact recognized by CONACYT. COFUPRO has fulfilled its commitment in terms of resource contributions, proposed improvements, and promoted efficient and opportune functioning. COFUPRO had learned that to avoid duplications in project financing, it was essential to cross-reference the RFP databases of all PFs. The Sectoral Fund adopted this practice, cross-referencing the databases for the PFs, the SAGARPA-CONACYT Sectoral Fund, and the basic science Sectoral Fund.

The improved use of information allowed a detailed analysis of which resources were supporting what, a better focus of the Sectoral Fund, and a clear justification for the themes included and excluded from the RFPs. Nevertheless, the Sectoral Fund's RFPs have not evolved in the same way as those of the PFs and still reflect a mechanistic view of research.

COFUPRO also learned from the Sectoral Fund. For example, researchers used to indicate to which institution they belonged. Now they must present an institutional endorsement when submitting the project, so that the institution and the researcher are jointly responsible for producing the results.

Another area of learning was project follow up. The Sectoral Fund still faces problems similar to those observed in the RFPs of the PFs: while the projects are being executed, controls are merely bureaucratic; there is no follow-up once projects are finished; and the information generated is not widely disseminated. Finally, the FAO evaluations were an important source of learning (see Section 5.5), but their contribution to the institutional learning has weakened as the methodology has failed to evolve and look for new areas of improvement.

6.3.3 Levels of Learning

Different levels and intensities of learning were observed. At the individual level, learning was initially focused on the PF presidents, followed by the managers. At the collective level, learning occurred in groups of managers. There was also learning at the organizational level, which, though not lacking some intensity, does face some obstacles. This learning is reflected in the more effective procedures adopted by the PFs as a whole for selecting projects, administering funds, and defining priorities. Learning is hampered, however, by strict rules of operation, inflexible supervision by SAGARPA, and

[3]For reasons of space, we have omitted mentioning many people who made important contributions to the foundations. Those mentioned are not the only ones who deserve recognition—they simply exemplify the importance of individuals in the institutional consolidation process.

lax governance mechanisms. Incipient learning processes at the innovation network level have been recently observed in some PFs.

One of the most serious problems for organizational learning is that the knowledge is located in individuals or groups (see Sections 3.1.2 and 6.2). Factors that limit organizational learning are as follows:

- The PFs have institutionalized certain methods and practices, but there is no standard procedures manual.
- There are varying degrees of adoption of good practices, as some PFs resist their adoption. COFUPRO operates by consensus: it has no formal mechanisms to pressure the foundations to fulfill agreements and adopt what it considers to be good practices. There are no rewards for those that do cooperate, nor is it documented how success is associated with good practices.
- Some boards are renewed gradually whereas others change suddenly. The changeover process does not in general allow preservation of the knowledge generated by the outgoing members.
- How to disseminate information is a lesson that has not been learned.
- There is little documentation of the learning process.

6.3.4 Capabilities Built

The PFs were able to build various capabilities that helped them improve the resource-allocation processes they administer. The biggest problem they face now is that, to continue improving, they need to develop new competencies. The capabilities already developed are described in this section.

Capacity to identify, assimilate, and improve new concepts and practices. The PFs and COFUPRO have often sought new concepts and practices, as exemplified by the identification of research demands and changes to the RFPs. Some individual foundations have experimented with new extension mechanisms. The problem the PFs face now is that in several relevant areas they have reached the international frontiers of knowledge, and as a group they lack centralized evaluation mechanisms, which limits the diffusion of successful experiences.

Capacity to influence research institutions. The PFs and COFUPRO are developing ways to influence the transformation of research institutions through participation in their councils, the definition of research priorities, and interactions with the researchers themselves. These interactions are helping the research institutions to be more sensitive to the needs of farmers (see Section 6.4). To develop such methods, COFUPRO has had to deepen its understanding of the nature of research and the new trends in science administration.

Capacity to learn how to learn. COFUPRO and several PFs are developing the ability to learn how to learn. They identify what they do not know and what they need to learn, search for help, and invest human and financial resources in learning. A clear example is the evolution in the methodology used to identify research needs (see Sections 5.3.5, 5.4.1, and 5.5). Nevertheless, learning is to a great extent still only present at the level of the individuals (especially the managers), and there are barriers to its socialization (for example, lack of suitable routines or management styles) that limit the consolidation of the capacity to learn how to learn.

Limited capacity to generate an alternative vision of science and innovation. The future of the PFs and COFUPRO depends on their capacity to generate a modern vision of science and innovation that strengthens their ability to innovate in all areas they oversee. An especially important responsibility is identifying emerging trends, whether they be trends in markets, scientific organization, methods of fostering innovation, or mechanisms for helping actors in the agricultural innovation system to actively position themselves in the face of these trends.

Traditionally, the vision of science held by the PFs was defined by a few people linked to COFUPRO (see Section 6.2.2). To deal with the highly complex issues COFPRO will face in moving from mostly managing the RFP to fostering innovation, the organization will have to develop its capacity to develop this vision.

6.3.5 Bias in the Learning Strategy

Learning may be intentional or an unanticipated result of actions taken with other ends in mind. If learning is to contribute to the construction of organizational capabilities, intentionality is required, that is, a learning strategy must be defined. This strategy should include two simultaneously active components: (1) operational (related to the acquisition, assimilation, and improvement of knowledge) and (2) strategic (related to innovation).

In its first years, COFUPRO implemented a learning strategy that simultaneously included the two components: the exercises to define the mission and vision and the adaptation of ISNAR's methodology illustrate this behavior. Subsequently the organization focused more on the operational component, and there were no clear efforts associated with the development of a new vision of research and innovation. In other words, there is no evidence of a discussion of where the foundations are, where they are going, and how they will get there.

6.3.6 Learning by Exploring

The PFs explored activities not contemplated in their original mandate. In fact, the activities most valued by the foundations were protracted processes in which the PFs catalyzed the emergence of innovation networks; these activities were not financed by the competitive funds (Muñoz 2005). These processes were ill defined in the sense that only the final goals were defined at the beginning, but there were no deadlines or intermediate milestones. This does not imply, however, that the processes were not monitored—in fact, what was monitored was the evolution and not predefined outputs. In these processes, the foundations used their resources to finance small, specific actions that removed hurdles as they emerged. Some examples of these processes are described in this section.

Analysis and consolidation of innovation networks in Michoacán. The 2004 FAO evaluation introduced the idea of innovation networks, that is, the notion that farmers innovate by interacting with different types of agents in the agricultural sector. In 2005 the University of Chapingo was contracted to conduct a study of innovation networks; the methods of Social Network Analysis were used to map technical, commercial, and social information flows among commercial lemon farmers in two of the state's valleys. The study continued with a technical and accounting analysis of lemon farms. From this information, each farmer was informed of how his or her productivity fared relative to the other farmers in the sample. If a farmer wanted to know about another farmer's methods (for example, how another farmer generated higher yields with a similar investment), the farmers were introduced to each other. The basic assumption of this project was that the farmers had a great deal of information, but the channels to share it did not exist. In the last stage, experts were hired to instruct the farmers and technicians on specific subjects related to their crops (plant nutrition, for example).

From this experience a course on management of innovation networks was developed. To date, eight innovation networks have been studied and 28 technicians have been trained in data collection and analysis of social networks and their strengthening. The objective was that technicians would eventually shift from being mere specialists in a product to become promoters of rural development. The project was jointly financed by the Michoacán PF and PRODESCA. Interestingly, the project was interrupted when the researchers could not make the transition from conducting an academic study and teaching extension agents how to use research tools to actually working with the farmers to strengthen existing social networks.

Heirs to the Land in Michoacán. The Heirs to the Land project seeks to link young people of the state's countryside to rural production and environmental conservation. Specifically, the project seeks to generate environmental management processes, participative research, and micro-enterprise development in different communities in the state.

In each community, the project begins with a participatory diagnosis of strengths, weaknesses, opportunities, and threats related to the environment, productive systems, and social participation. Children, young people, and external facilitators participate in these diagnostic exercises, in which the young participants use secondary sources of information, fieldtrips, and interviews with key actors from their communities. These activities end with the participants documenting their experiences and receiving training in the analysis of different production systems, preparation of business plans, and the importance of conserving natural resources. The training includes visits to projects-in-progress and hands-on workshops on sustainable projects (for example, composting). An important objective of the project is the preservation of family histories and traditional knowledge.

Northwest Technical Consortium in Nuevo Leon. The Northwest Technical Consortium (CTN) is a regional organization created in 1997 with the participation of the Universities of Tamaulipas, Nuevo Leon, and Coahuila, INIFAP, the Agriculture-Related Trusts of the Central Bank of Mexico (FIRA), cattle ranchers associations, SAGARPA, and the governments of the three abovementioned states.

The number of researchers and technicians who participate in the CTN varies, with about 10 or 12 collaborating permanently and another 30 participating sporadically. The CTN primarily focuses on technology transfer, farmer training, and provision of services rather than on research. It caters to medium-sized cattle ranchers.

Most of the projects submitted by the CTN to the foundation involve adapting and transferring technologies from domestic and foreign institutions. Most of the CTN's research is in the form of individual projects financed by the universities and CONACYT. The Nuevo Leon PF primarily supports extension events. In practice, the CTN works as a facilitator linking researchers and university students with farmers. The main problems the CTN confronts are (1) lack of incentives for hiring technicians and researchers, (2) shortage of financial resources, and (3) aging researchers.

The CTN Promotional Committee is a laboratory that offers services and undertakes research in the areas of animal health, diagnosis, geographic engineering, digitalization, and wildlife. The Promotional Committee has agreements with INIFAP and the University of Nuevo Leon. The CTN submits projects to CONACYT and to the Mixed Funds.

GGAVATT in Nuevo Leon. The GGAVATT in the state mainly operate among goat and pig farmers. These groups receive resources from the Alliance for the Countryside and do not have a formal relationship with the Nuevo Leon PF. Nevertheless the interactions that occur in these groups between farmers and researchers have allowed the identification of potential problems and the development of projects submitted to the PF. They have also facilitated the PF's establishment of technological validation plots and contributed to the adaptation and diffusion of technologies among farmers.

Papaya maradol in Puebla. In 2002 the Puebla PF, FIRA, the 32nd Agricultural Technological Institute of Tecomatlán, and the Institute for Fundamental Research in Tropical Agriculture, Havana, signed an agreement for the transfer of technology for the production and postharvest management of papaya maradol (a variety of papaya).

After conciliating antagonistic interests, several producers of papaya maradol organized themselves with a collective vision of establishing irrigated cultivation of papaya maradol and transferring the technology for farming, postharvesting, transformation, and commercialization of the fruit. The project covers nine localities in the Puebla Mixtec region. Two of the most important problems in the organization were political differences among the municipalities and a culture developed by farmers of organizing to receive public subsidies. With the support offered by the Puebla PF, it was agreed that the technical-commercial organization was to be separate from the political organizations. After decades of patronage, many farmers organize themselves only to receive support from the government. Initially several farmers in the project looked for this type of support, but left as their expectations were not fulfilled. Today only those who understood the advantages of receiving training, support for technological missions, and information from highly qualified technicians remain. Currently the project covers 120 ha, has trained 107 farmers and 43 technicians, and has significantly increased profitability per hectare.

Jamaica production and processing in Puebla. In 2001, the Puebla PF started to work with Jamaica farmers in the Mixtec region.[4] The project's initial objective was to foster the production and processing of Jamaica flowers for fresh consumption and for its extracts, marketing them in the domestic market. With time, the objective expanded to include the manufacturing of the flowers into industrial products (concentrates and extracts, bottled water, jam, and liquor) and organic cultivation.

The Puebla PF is contributing to organize the Jamaica industry in the state, in particular, strengthening its managerial capabilities. To this end, the farmers established a company that hired four young professionals: an agronomist as field manager, a biochemist as manager of the processing plant, a business-trained sales manager, and a general manager. These four professionals are in charge of supervising the processing and marketing of the flowers, while the farmers serve on the board of directors.

Two additional projects related to Jamaica were organized for the farmers' wives and children. The former were taught to cook meals with the plant, and the children were taught to feed the seeds to hens for egg production. The Puebla PF financed the investment in hens. The Jamaica project is 5 years old, and some decisionmakers believe that it is time for the PF to abandon it.

Industrial firm CAIVO in Puebla. CAIVO is a private firm that was created with an investment of more than 15 million pesos (equivalent to about US$1.35 million). The business includes a slaughterhouse authorized to export to the United States, preparation of lamb fine cuts, and the sale of vacuum-packed lamb *barbacoa* and *mixote* (traditional Mexican foods) targeted at the Hispanic market in the United States. At the moment CAIVO has 450 sheepherder partners, of which 80 operate independently and 370 are organized in four small farmer societies. To sell through

[4]Jamaica is the flower of the *Hibiscus sabdariffa* L. In Mexico it is processed into several products, such as drinks, jams, and creams.

CAIVO, farmers have to buy shares in it. The current price of a share is 50,000 pesos (equivalent to US$4,500), and each share grants the right to slaughter 200 sheep a year.

The project began in 2002, when the Puebla PF supported the project "Prospective Analysis of the Sheep Agro-Food Chain in the State of Puebla." Based on this project, it was decided to set up a high-tech slaughterhouse specializing in sheep. The PF's main contribution was to help in the organization of the lamb farmers and in the preparation of a business plan.

CAIVO operates a mobile technology transfer unit that specializes in raising and fattening lambs; the unit was financed by the foundation. The PF also financed a laboratory for the slaughterhouse and supported commercial promotion. The financial support that the PF offers to CAIVO is not significant, but it helps organize the farmers and supports research, training, and technological missions. For example, the Puebla PF contracted two Canadian specialists to instruct operators in new techniques of meat cutting.

6.3.7 Impact of Cultural Duality on Learning Processes and Capacity Building

As discussed in Sections 6.1 and 6.2, although there is a set of shared values in the PFs and COFUPRO, there are also two subcultures: one associated with the managers and the other with the presidents. These subcultures influence how individuals learn, how they socialize that learning, and how the organization absorbs what is learned.

The managers' culture is focused on technical activities and is structured according to an organic model in which each manager participates in a workgroup responsible for improving specific types of activities. These activities result in learning processes that are socialized in the group and generate the conditions for organizational learning and gradual acceptance at the level of COFUPRO. Transmission of the culture to new members, generation of new routines, and socialization of learning depend on the culture-bearing managers staying in their posts.

In contrast, the presidential culture is focused on building external relations useful to the foundations. To do this, the presidents mainly use informal channels based on their proximity to power. Given the nature of their activities, their learning occurs basically at the individual level and is hard to socialize. Because the skills to develop these relations are a function of their personal reputations as farmers rather than belonging to the foundation, this culture cannot be passed to managers but is also nontransferable to future presidents.

6.4 Impact of the PFs

It is not possible to unequivocally identify all effects of the PFs, because they were only one element of a set of actors participating in a process of change triggered by the federal government. This section analyzes some of the identifiable influences the PFs had on the agricultural research and innovation systems, including some qualitative assessment of their impact on poverty. It should, however, be emphasized that the evidence is partial and the processes complex, so it is not possible to identify and quantify all effects or allocate the effects among the different causes.[5]

It is also difficult to identify the effects on the extension system, because its changes were more profound than those experienced by the research system. After the dissolution of the public extension institutions, various private and civil society agents organized lim-

[5]When several causes interact in a process, the effects belong to the whole set and cannot be apportioned among individual causes (Axelrod and Cohen 1999; Ragin 2000).

ited extension activities. But these experiences have not been analyzed in depth. The most notable effort to set up extension mechanisms was PRODESCA, the program that sought to foster the emergence of a market for technical services. The FAO evaluations and the fieldwork conducted during this research indicate that this program has had numerous operational problems. The PFs are also exploring mechanisms to disseminate technical information to farmers and foster innovation; some of the mechanisms explored are very innovative, as the one that uses Social Network Analysis to improve information flows among farmers (see Section 6.3.6). Although the PFs devote a significant share of their resources to extension activities, there is no centralized mechanism to either evaluate them or learn from the different tools they use. This deficiency has been recognized by the PFs, and they have launched an initiative to overcome it (see Chapter 8).

As explained in Section 5.2, the PFs were created to induce changes in the agricultural research system to make it more responsive to the needs of farmers. The nature of these changes (and consequently the influence of the PFs) fluctuated with changes in personnel, lessons learned, and changes in the political system. Transformation of the research system was triggered by the crisis of confidence in INIFAP in the early 1990s. At that time, a relatively limited reform of the institute was attempted. This effort essentially consisted of establishing new ways of interacting with farmers (through the PFs) and securing new resources for operational funds (see Section 5.1).

Beginning in 2000, an attempt was made to accelerate INIFAP's renovation by transforming it from an organization completely dependent on SAGARPA to a Public Research Center. Nevertheless the pace of change was slow because of the difficulties of transforming an institution without adequate tools—especially shrinking budgets and inflexible incentives and hiring policies. As a response to the sluggishness of change, in 2003 SAGARPA commissioned the Inter-American Institute for Cooperation on Agriculture to carry out an evaluation of INIFAP by an international team.[6] Implementation of the commission's recommendations began in 2004, but progress continues slowly for the abovementioned reasons.

In 2003, the Colegio de Posgraduados hired Wageningen University to evaluate all its activities, especially its research and programs for graduate studies. The difference between this evaluation and the abovementioned one is that INIFAP's was commissioned by SAGARPA, whereas the Colegio's was contracted by the evaluated institution itself. Reform of the other agricultural research institutions (including most of the universities) is being implemented even more slowly. Because universities are isolated from external influences by their autonomous regimes, changes follow the internal dynamics of each institution.

Before the creation of the PFs, INIFAP followed SAGARPA's directives; internally, the researchers reported to the institutional hierarchy. The PFs weakened these lines of authority as they awarded resources directly to the researchers. Owners of their own funds, researchers dedicated less time to resource-starved institutional projects. In this way, the PFs started to influence the research agenda. The changes to this agenda were in general limited by the difficulty in getting the researchers to adequately respond to the relatively broad RFPs (see Sections 5.5 and 6.3.2). The lack of response was what motivated the change in the mechanisms used for the RFPs.

Another significant consequence of the first interactions with the PFs was that the researchers now had a direct channel of com-

---

[6]Evaluations by international experts are a common practice in the universities of many developed countries and the CGIAR centers.

munication with stakeholders in their state's agriculture, which further weakened the centralism of the national research institutions. This process became even more pronounced with the strengthening of INIFAP's regional offices.

Recognizing local political interests and the needs of research institutions, the PFs often gave preference to state researchers when evaluating proposals, even though they were often not the best in their field. The preference for local researchers was more pronounced in the early years of the PFs; as the weaknesses of the local teams became more apparent, some PFs began inviting out-of-state researchers to take part in the RFPs. Foreign professionals were also contracted for specific research projects. Nevertheless, in spite of some isolated cases in northeastern Mexico, these contracts are still not a commonly used tool to strengthen domestic teams. Although the influx of out-of-state researchers has somewhat improved the research financed by the PFs, it has not strengthened state research institutions.

The PFs organized several courses on the preparation of research proposals and the use of the log frame, but these courses did not improve the quality of proposals. The reason is that quality depends essentially on the researcher's capability, and the latter does not depend on mastering formal instruments for project preparation but rather on the researcher's excellence, his or her interactions with the international academic community, institutional cultures, and the incentives offered.

Neither did the mechanisms used for selection of proposals at the state level help to strengthen the research system. Each PF set up its own technical committee; its members were often selected as representatives of their state institutions rather than for their professional quality. In addition, the smaller size of the state research systems did not facilitate transparency (Ekboir 2004).[7] These factors damaged the credibility of the proposal selection mechanisms at the state level.

Despite these problems, the quality of the projects as a whole improved, because the PFs gradually identified the best researchers and used this information for decisionmaking. It should, however, be stressed that this improvement is not a consequence of the competitive grants, but rather of the use of reputations in the selection of proposals. These reputations are built through stable interactions between the PFs and researchers. Basically, any stable, repetitive mechanism that allows reputations to be established would have the same effect.

An important impact on the research system was the opening up of finance sources for institutions that traditionally had problems accessing research funding. Major beneficiaries were the Colegio de Posgraduados and the Universidad Autónoma de Chapingo. The state universities also benefited, though their competitiveness was limited by their weak research capabilities.[8]

The PFs had effects unforeseen at their creation. As mentioned in Sections 5.3 and 5.4, COFUPRO's presidents had good access to the secretary of agriculture and submitted proposals for agricultural policies and the transformation of the research system. The

[7]The lack of transparency does not imply that they acted in bad faith. The transparency of the process depends on it being "blind"; that is, the researcher should not know who evaluated the proposal. In a small system there are very few researchers in each subject area. Thus a researcher may be able to identify who evaluated a proposal just from the nature of the comments and the writing style.

[8]Except for some traditional universities, such as the Autonomous Metropolitan University (UAM) or the National Autonomous University of Mexico (UNAM), the pressure on university professors to conduct research is relatively recent. But most professors have neither the training nor the incentives to do it. Despite this general weakness, a few strong research teams have emerged in some universities.

PFs also introduced the idea that innovations are developed and diffused by networks. SAGARPA adopted this idea in 2006 for the evaluation of the Subprogram for Research and Technology Transfer (SITT); INIFAP is also beginning to set up internal research networks by products, but without a clear vision of how the networks should link with external stakeholders, what their potential is to promote innovation, or what actions are most effective to further this aim. The problem is that strengthening agricultural innovation networks is a relatively new instrument internationally, so there are few experiences from which Mexico can learn.[9]

Members of COFUPRO and the PFs are on the board of directors of INIFAP, the Colegio de Posgraduados, and other research and teaching institutions. They are often the only nongovernment representatives on these boards, and their participation is important, because their nonpublic status gives them more freedom than other board members have to question the public-sector representatives.

The presence of COFUPRO in these governing bodies reflects a new dynamic in the research system. At the beginning of the 1990s, INIFAP was the de facto coordinator of the agricultural research system and as such was influential in the running of the PFs. Today COFUPRO occupies this central position (distributing operating funds and generating data useful for the prioritization of programs and policies) and has a strong influence on the management of research institutions. It should be noted, though, that COFUPRO does not play as pivotal a role as did INIFAP up until the 1990s, because other institutions also finance agriculture-related research projects.

The forums in which research needs are determined and the technical committees in which proposals are evaluated put many researchers in contact with farmers and researchers from other institutions. New patterns of interaction among actors in the innovation system emerged from these contacts. During the field work, several researchers commented that, based on these relations, their own priority setting and research methods have changed. These changes, however, resulted from the researchers' personal motivation and not from changes in the institutional cultures. To have a bigger impact, the PFs will have to explore new interventions to transform the research system (see Chapter 8).

COFUPRO was an active partner in the SAGARPA-CONACYT Sectoral Fund until 2006, when COFUPRO withdrew from the fund, because it believed that the fund was too bureaucratic and was not generating useful information. CONACYT's policy in all Sectoral Funds has been to let the partners define the priorities. In the case of the SAGARPA fund, the priorities were mainly defined by COFUPRO, which also made sure there were no overlaps between the RFPs of the states and the Sectoral Fund. COFUPRO was also an active member of the fund's evaluation commission. There was concern among certain actors that the needs identified by COFUPRO might not represent all sector needs, particularly those of small farmers or products not included in the prioritized chains. Currently COFUPRO participates in the fund's technical committee, but since 2008 the operational rules do not allow it to make financial contributions to the fund.

COFUPRO was influential in the Sectoral Fund because it demonstrated a commitment to permanent improvement that is not replicated in other funds, and it contributed a methodology to identify demands that satisfied

[9]Strengthening of agrifood chains has received a great deal of attention from decisionmakers and researchers. However, this approach is more specific than support of innovation networks.

the other partners. The technical committee's working group met monthly at alternating venues; the meetings were active and forced participants to make decisions. COFUPRO does not participate in any other agriculture-related Sectoral Funds, nor do the PFs participate in their states' Mixed Funds, suggesting a lack of coordination with nontraditional actors linked to the sector.

The two biggest gaps in the ongoing transformation of the innovation system in general and the PFs in particular are the lack of a debate about what the agricultural innovation policies should be and which extension mechanisms are best for the needs of different types of farmers.

6.5 Role of PFs in Poverty Alleviation

Agriculture can contribute to poverty alleviation in two ways: the expansion of commercial agriculture creates employment and reduces food prices, and some small farmers can become commercial farmers in higher value markets (see Section 3.4). Although the PFs can contribute to poverty alleviation in both ways, its main contribution may result from the exploration of new instruments to help small farmers enter more profitable value chains and to increase researchers' participation in innovation processes (see Section 6.4).

Until now, the PFs have mainly supported commercial farmers, but there are indications that this focus is changing. Several projects have already had important economic impacts on commercial agriculture, but their effects on poverty have not been evaluated. The most innovative PFs have also implemented some projects that helped small farmers integrate into commercial value chains. These projects, however, were few and only involved a limited number of farmers. Three important deficiencies of these projects have been identified. First, the PFs have not developed methods to identify innovative farmers (see Section 3.4); participation in the projects was initially broad, but eventually only farmers with the strongest commercial and innovative capabilities remained. The consequences of this ex post selection were that many of those who failed felt frustrated, there was a great deal of tension in the groups during the process of culling people who did not participate in the collective effort, and resources were wasted in dealing with these failures.

Second, the PFs have not been able to learn how to scale up successful local projects or how to replicate them. And third, the lessons learned by the individuals who managed the projects were not shared with other PFs, hampering collective learning. In other words, the lessons learned by individuals did not become shared knowledge. This deficiency has been noticed by some foundations, and the creation of a specialized structure to facilitate learning has been discussed. The structure would combine decentralized experimentation with centralized learning. Thus it would take advantage of the creativity of the individual PFs and establish a small team of experts in innovation systems to assess the experiments, suggest new instruments to be tried using pilot projects, and disseminate the lessons learned among all PFs.

The PFs are also actively inducing their presidents and managers to commit to poverty alleviation. In a recent survey of presidents and managers, the majority of respondents mentioned that people in these positions should have a very strong sense of duty and a commitment to poverty alleviation. When asked to identify one successful project funded by their PFs, they all mentioned multiyear complex projects that involved several actors (examples of such projects can be found in Section 6.3.6). Interestingly, nobody mentioned projects that had been financed through the traditional RFPs, even though some projects had important outcomes. Despite their commitment, few presidents or managers had a clear idea of how to implement pro-poor programs.

It is expected that the PFs will accord poverty alleviation higher priority in the future, because the interactions between the presidents and managers from different PFs work as a filter that marginalizes people who do not share these values, especially among the presidents. The PFs are also planning to set up a program to build the capabilities of presidents and managers to manage agricultural innovation programs; the program will include a discussion of poverty alleviation. Finally, the PFs are engaging researchers to participate in projects that target poverty, effectively inducing them to change their research routines and facilitating farmers' access to scientific information.

CHAPTER 7

Conclusions

The complexity of agriculture in developing countries has increased in recent decades. New products, markets, and actors have emerged, creating new opportunities for innovation. Responding to these opportunities, the goals of development policies shifted from increasing local food production and addressing market failures to creating competitive advantages, increasing agriculture's sustainability, and reducing the growing inequality that resulted from globalization. A crucial factor in addressing these issues is how to create the individual and collective capabilities all actors in the agricultural innovation system need to help farmers integrate into dynamic markets (as entrepreneurs or employees).

The study of the nature of innovative capabilities in private firms has attracted the attention of researchers since the late 1980. In contrast, the nature of innovative capabilities among small farmers, urban poor, NGOs, and the public sector has barely been analyzed. This report contributes to filling this gap by answering two questions: How can an organization managing public funds for research and extension sustain organizational innovation over extended periods? And how can it learn and adapt to maximize its impact on the agricultural innovation system? This report also briefly explores how innovation programs can be used in poverty-alleviation strategies. The answers were researched by assessing the institutional dynamics of the Mexican PFs, a highly innovative civil society organization that manages public funds for research, extension, and innovation.

The main finding was that the innovativeness of the PFs emerged from the interaction among internal and external factors. The external factors were expanding profitable markets for agricultural products; a dynamic private sector searching for new technical, commercial, and organizational innovations to take advantage of these emerging opportunities; and a public sector that allowed the PFs to evolve mostly in response to their internal dynamics and the needs of farmers. The internal factors were the presence of highly innovative and committed individuals who questioned their own capabilities, a decentralized governance structure that allowed experimentation by individual PFs, and a centralized structure that fostered collective learning. These factors resulted in strong collective learning capabilities, especially the ability to identify knowledge gaps, define strategies to fill them, and use the new knowledge to explore new activities and organizational structures. The learning capabilities were not distributed evenly over all PFs but instead were concentrated in a few of them.

The study of the evolution of the PFs yields important lessons for the management of research and innovation in developing countries. First, because of the complexity of innovation processes, science, extension, and innovation policies should be flexible and evolve as new information becomes available and new capabilities are acquired. The flexibility should be implemented at all levels. In the Mexican case, the federal and state governments allowed the PFs to change, although it was not in the Mexican political tradition to do so. Some PFs developed new capabilities by implementing actions that were not in their initial mandate. The flexibility should also be reflected in changes in the goals and instruments of policies. Initially the PFs were mandated to finance research and extension, but they started to focus on innovation as the limitations of the original mandate became apparent. In a similar way, new financing mechanisms were tried when the researchers and technical advisers did not respond as expected to competitive grants. The flexibility was not chaotic (in the sense that the PFs were restricted in their exploration) but was managed through an active dialogue between COFUPRO and SAGARPA.

Second, innovative abilities have a much skewed distribution. In the PF less than 10 percent of presidents and managers were innovative. Similar distributions have been found in other research areas such as education and psychology. The uneven distribution should be explicitly considered in the appointment of project managers and in the design of programs seeking to increase poor households' agricultural income.

Third, the development of innovative capabilities depends on strong and sustained commitment by the authorities (whether in the public sector or in a private firm). The innovative presidents and managers made major personal efforts to explore alternatives and share their lessons with other PFs. The state and federal authorities initially tolerated the changes and later encouraged them. Interestingly, they did not attempt to strengthen capabilities in the public sector.

Fourth, the policies should not be left to evolve randomly but should be managed by balancing the exploration of new instruments with the exploitation of those that have shown their efficacy. Balancing these two strategies requires an effective search mechanism that should combine decentralized exploration with centralized learning. Overreliance on centralized exploration can miss important opportunities, whereas decentralized learning alone hampers the sharing of useful information; decentralized exploration can be complemented with directed searches when the opportunities or needs are clear. The directed searches should not attempt to establish major programs from the start but rather implement pilot projects to test the assumptions used to design the program. Only after the pilot programs have shown their effectiveness should they be scaled up. Action research to test different types of interactions in the innovation networks should be the basic approach to design the pilot programs.

Exploration should not be restricted to financing traditional research. It should also include the analysis of the joint dynamics of innovation and poverty and the identification of new ways to foster the emergence of innovation networks and involve new actors in innovation processes. A final important element in the search strategy is a monitoring system to guide the exploration. The two most important factors that determine the system's efficacy are the flexibility of the monitoring system and the set of indicators to be monitored. A key feature of complex systems is that interventions often have unexpected results. A monitoring system that constrains itself to a predetermined set of indicators would miss those results. In the case of the PFs, the evaluations conducted by FAO focused on the alignment between the identified demands and the projects financed. This narrow criterion missed the most important impacts on the innovation and research systems.

The indicators should provide timely and accurate information on the evolution of the process. For example, measuring the benefits or the rate of return of research projects does not provide information useful for managing research, because (1) the impacts do not result just from research but from the interaction of several factors, including alternative technologies, other sources of information, markets, policies, and the strategies implemented by competitors; (2) quantitative ex post impacts do not provide information on the factors that contributed to the outcome; and (3) even if those factors could be identified, it is unlikely that they will have the same influence in the future. The new literature on management based on complexity theories indicates that when the processes are complex and the causality between the inputs and outputs is not simple and clear, monitoring should not focus on outcomes but on the process and the contributions made by each actor.

The set of independent PFs became an effective exploration structure. However, they did not develop an effective learning mechanism. COFUPRO served as a means to centralize and diffuse the lessons learned by the individual PFs, but it was up to the latter to decide whether to share the lessons they learned and use the instruments developed by other PFs.

Fifth, individuals play a major role in the success or failure of innovation processes and policies. Traditionally, policymakers have paid a great deal of attention to the design of the organizations that will implement the programs and the rules they must follow. The selection of capable individuals to run the organizations, however, has received less attention. Program directors are usually selected not for their capabilities but for their allegiance to the decisionmakers. The Mexican experience clearly showed the importance of individuals who challenged the boundaries and explored new activities. Selection of capable individuals is a necessary but not sufficient condition for success. A continuous training program to build their human and social capital is necessary to maintain innovativeness in the long run. The analysis of the PFs also showed that capable and committed individuals are crucial for the success of poverty-alleviation programs.

Sixth, the programs should have effective governance structures in which innovators can influence decisions. For example, although farmers have participated on the boards of many public agricultural research institutes, few have had real impact on how these institutions operate.

Finally, innovation is more than just applied research. To effectively participate in innovation processes, public research and extension institutions must adopt a new conceptual framework in which they recognize that they are not the central actors but play an important supporting role. Adoption of this new framework leads to new operational routines in which researchers integrate into innovation networks.

The experience of the PFs clearly showed the importance of capable individuals and flexible structures for the implementation of effective policies in rapidly changing environments. This focus clearly contrasts with the traditional approaches that emphasized careful planning and rigid execution. This new approach imposes major demands on the public sectors of developing countries, which usually have weak public institutions and organizational cultures that hamper innovation. These two factors are also important for the design and implementation of poverty-alleviation programs. Because contractual arrangements between the funders and the operating agent are usually rigid and agents on the ground do not have the time and resources to reflect on their actions, these programs generally do not change once implementation starts. Given the complexity of poverty, however, it is most likely that the original assumptions used in designing the program will not be completely valid as the program progresses. This rigidity should be avoided, allowing flexibility in the implementation and setting specialized units to experiment across several projects, similar to the one that has been created by key humanitarian organizations.[1]

[1]This unit is called the Active Learning Network for Accountability and Performance in Humanitarian Action (ALNAP). This network, which is based in the United Kingdom, incorporates many key organizations and experts from across the humanitarian sector. ALNAP includes members from donor, NGO, Red Cross/Crescent, UN, and independent and/or academic organizations and is dedicated to improving humanitarian performance through increased learning and accountability. As a result, the network is able to use the broad range of experience and expertise within its membership to produce tools and analyses relevant and accessible to the humanitarian sector as a whole. Further information on ALNAP is available at www.alnap.org.

CHAPTER 8

Epilogue

Our original report was submitted to COFUPRO in August 2006. During the second half of that year, Mexico's new president consolidated its power after a highly contested election and a narrow win. The troubled political transition in several secretaries of the federal government, including Agriculture, lasted for most of 2007. Late that year, SAGARPA substantially changed the rules that governed the PF operations, eliminating the guaranteed funding for the PFs and forcing them to compete with other organizations willing to implement research and extension projects. The result was a more discretional use of resources by SAGARPA. The foundations contribution to COFUPRO was also eliminated from the regulations, forcing COFUPRO to seek alternative funding. These changes jolted the PFs, because they threatened the latter's existence, especially that of the weaker PFs. After strong lobbying efforts and the involvement of several presidents, governors, and other actors in the agricultural innovation system, the operational rules were changed again in 2008. This time, the PFs were recognized as the administrators of the funds for agricultural research and extension. The PFs could not convince SAGARPA to reinstate the mandatory contribution to COFUPRO, however. To finance itself, COFUPRO had to seek special funding from SAGARPA for a project to monitor the PFs. The result has been a weakening of the relationship between COFUPRO and the PFs. If this trend continues, COFUPRO may change from a learning and lobbying tool of the PFs into an instrument used by SAGARPA to control the PFs. The PFs are conscious of this threat and are currently seeking new mechanisms to fund COFUPRO. The decentralized decisionmaking process is hampering agreement on a permanent solution, however.

In addition to the funding issues, a new relationship developed between SAGARPA and COFUPRO. The interactions are now less dependent on the personal relationship between the secretary and COFUPRO's president; in other words, it is both more distant than before and more institutionalized. The lack of support from SAGARPA has deprived COFUPRO of an important tool to foster collective learning.

At the same time, the relations between some presidents and managers with policymakers at the federal and state levels have intensified, increasing the latter's contributions to the design and implementation of agricultural policies. COFUPRO has also increased its collaboration with other stakeholders in the agricultural sector, including the National Association of Secretaries of Rural Development and the National Association of Managers of Applied Research and Technological Development, CONACYT, and farmers' associations.

These changes have introduced new dynamics to the PFs and COFUPRO, but it is still too early to know how these changes will unfold. Despite this uncertainty, COFUPRO has started to implement some of the recommendations made in our original report with the explicit objective of strengthening the innovative capabilities of the PFs. Among the most important actions being implemented are the following:

- Creation of a training program for managers that will include formal and informal actions. Negotiations are under way with UAM to create a set of courses on management of agricultural research and innovation. After completion of the courses, students will be allowed to write a thesis and receive a master's degree. A tutoring program will also be created, in which the best managers will advise newly appointed managers, and managers will be encouraged to visit other PFs to see firsthand how they operate.
- Creation of a training program for presidents and board members. This program will also include formal and informal instruments.
- Creation of a commission to analyze the governance of the set of PFs and COFUPRO. This commission used our original report and several workshops to prepare recommendations on how to improve the interactions among the different actors that participate in the PFs and how to strengthen collective learning. The PFs are starting to implement the recommendations.
- Creation of a specialized team to advise the PFs on institutional innovation and on how to manage agricultural innovation.
- Establishment of a research project to assess learning by the individual foundations and the impact the PFs are having on the research and innovation systems. The project is financed by the Sectoral Fund operated by SAGARPA and CONACYT. It will collect information on the projects the PFs operate (or finance with funds other than those from the Alliance for the Countryside), on how researchers have changed their working routines in response to the different activities of the PFs, and on the interactions researchers are having with other actors in the agricultural innovation system.
- Assessment of innovative practices. COFURPO organized a workshop with managers to identify innovative practices they developed in their foundations. This exercise is a first step toward the creation of a structure to foster centralized learning.
- Creation of new spaces for the interaction among researchers and farmers. A few foundations are experimenting with new methods to foster the integration of INIFAP researchers into innovation networks, including new incentives and financial arrangements.

APPENDIX A

Organizations Discussed in This Report

See the list of acronyms near the front of the report for expansion and English translations of the organizations listed here. Unless otherwise stated, these organizations are based in Mexico.

CIAT is one of the 15 international agricultural research centers that belong to the CGIAR. It is based in Cali, Colombia.

COFUPRO is an office created by the PFs as a whole that (1) represents them in dealings with the federal government, (2) coordinates collective learning, and (3) facilitates the creation of an institutional culture.

CONACYT is the national research council. Its main activity is the administration of public funds for research and innovation.

CREAs are small groups of farmers that organize themselves to validate and diffuse technologies. The individual groups are organized in a national association. CREAs operate in Argentina.

CTN is a consortium of universities, research institutes, financing institutions, and farmers' organizations from northwestern Mexico. CTN finances and conducts research and extension activities.

FAO is part of the United Nations. It leads efforts to eradicate hunger and is a neutral forum where member states meet to debate agreements and policies. It also provides technical cooperation to member states.

FIRA is a trust fund created by the Mexican government in the central bank of Mexico to finance investments, capacity building, technical assistance, and technology transfer in the agricultural, livestock, forestry, and fisheries sectors.

GGAVATT are small groups of farmers assisted by a technician that validate cattle production technologies and then share their findings with similar groups.

ICAMEX is a research center financed by the State of Mexico that works on agriculture, livestock, forestry, and aquaculture.

IICA is an international technical cooperation organization that is part of the Organization of American States.

INIFAP is the national research institute specialized in agricultural, livestock, and forestry.

ISNAR was a center that belonged to the CGIAR system. In 2004, some of its activities were absorbed by IFPRI.

PFs are civil society organizations that manage public funds to finance research, extension, and innovation projects.

PIEAs are farmers' associations that finance agricultural research. They usually specialize in crops, livestock, or forestry. Most PIEAs exist only on paper and do not mobilize resources.

PRODESCA is a federal program that finances the provision of technical advice to farmers.

SAGARPA is the federal ministry of agriculture, livestock, forestry, and fisheries.

SITT is a program jointly financed by SAGARPA and state governments that finances research, extension, and innovation projects. SITT is operated by the PFs.

SNITT is the national system for the transfer of research and technology. It is a formal structure mandated with coordinating all public research and extension institutions. Private institutions may participate voluntarily.

UAM and UNAM are Mexican public universities.

Figure A.1 Relationships among the main institutions discussed in this report

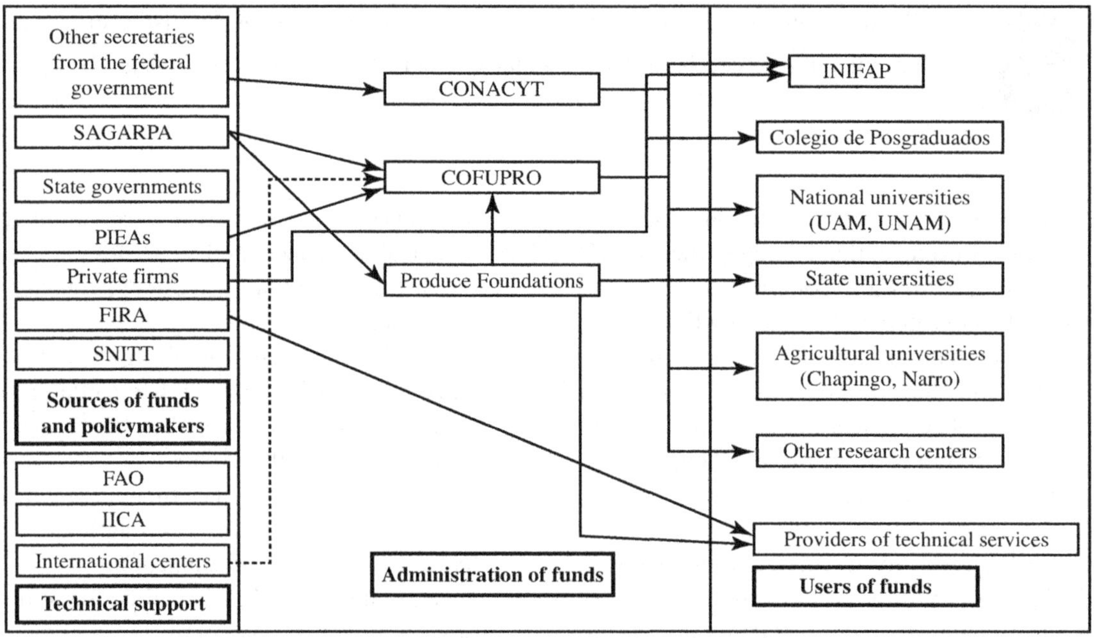

Note: COFUPRO, National Coordinator for the Produce Foundations; CONACYT, National Council for Science and Technology; FAO, Food and Agriculture Organization of the United Nations; FIRA, Agriculture-Related Trusts of the Central Bank of Mexico; IICA, Inter-American Institute for Cooperation on Agriculture; INIFAP, National Institute for Forestry, Agricultural and Livestock Research; PIEA, Foundation for Agricultural Research and Experimentation; SAGARPA, Secretary of Agriculture, Livestock, Rural Development, Fisheries, and Food; SNITT, National Research and Technology Transfer System; UAM, Autonomous Metropolitan University; UNAM, National Autonomous University of Mexico.

APPENDIX **B**

Alliance for the Countryside

The program known as Alliance for the Countryside (Alianza para el Campo) began operations in 1996. Its objective was to support the modernization of agriculture to enable it to better compete in the North American Free Trade Agreement. The Alliance is a set of agreements between the federal government on the one side and state governments and farmers on the other. An agreement was signed in each state.

The Alliance has three main programs: farming support, livestock support, and rural development. Additionally, the Alliance has several subprograms: plant and animal health; research and technology transfer; food safety; aquaculture and fishing; the national information system for sustainable rural development; and the stabilization, support, and reorganization program for coffee cultivation.

The Alliance's public funds come from the federal and state governments; farmers contribute in different ways according to the rules of each program. The farming and livestock support programs finance a share of farmers' investments in physical capital, and the farmers themselves provide the rest. The research and technology transfer subprogram is administered by each state's PF, which asks research institutions, technical advisers, and farmer groups to submit proposals for projects they want to implement in the state.

APPENDIX **C**

Recommendations for Strengthening the Innovative Capabilities of the PFs

The PFs were an important institutional innovation in the Mexican agricultural sector. Nevertheless, there are indications that the model followed by the PFs during its 10-year life may be losing creative potential because of the solidification of their structures and the maturing of the areas in which their impact has been greatest (identification of research demands, opening communications channels among innovation system actors, and improving efficiency in the RFP process as well as in project administration and follow up). In the past, the capacity to contribute ideas and structures for the implementation of scientific and technology policies was derived from the learning mechanisms and from the governance structures developed by each PF and by the whole set. The PFs will be able to maintain their creative dynamics by developing new learning capabilities, especially in the design and implementation of innovation policies and programs, in support of the transformation of research institutions, and in the emergence of effective extension structures (see Section 3.2).

This appendix presents recommendations for strengthening the operations of the PFs. Effects on agricultural productivity and competitiveness, natural resource management, and poverty were not investigated, as they are beyond the scope of the present study.

C.1 Governance

C.1.1 Relationship with Other Actors in the Agricultural Innovation System, including SAGARPA and State Governments

- It is recommended that each foundation establish or strengthen its interaction with other administrators of research funds (for example, the state science and technology councils), with the aim of consolidating its legitimacy, improving the effectiveness of its actions, and catalyzing local debate on the role of science in state development.

The PFs always had direct access to the official circles in which agricultural policy is formulated, whether in SAGARPA or in most states. These circles valued the information the PFs generated, the channels of communication they opened with apolitical commercial farmers, and their flexibility in the administration of funds. COFUPRO's contributions to the SAGARPA-CONACYT Sectoral Fund also consolidated the relationship with CONACYT.

Both federal and state authorities and the hierarchies in the public agricultural research institutions recognized the PFs and COFUPRO as the voice of apolitical commercial farmers. It is important that the PFs recognize that their acceptance stems in part from their legitimation by the authorities and from a group of researchers. The legitimacy that comes from participation in power structures is similar to that of CONACYT, which is not representative but is well known by the academic community. The recognition legitimized the PFs but at the same time caused certain actors that were not included in its structures to question its representation and behavior.

To boost legitimacy, the PFs will have to continue contributing to the formulation of agricultural policy and collaborating with other actors in the research system. Their interactions with COFUPRO and CONACYT contributed to this process, but the PFs have not interacted with the same intensity with the state science and technology councils, the Mixed Funds, or those universities that have not traditionally been part of the agricultural research process but which could contribute to the objectives of the PF (for example, UNAM or UAM). Although the lack of interaction probably has its origins in institutional inertia on both sides, the PFs need to recognize that they are part of a system that is broader than just agricultural research and that they should consequently strengthen their ties with the system as a whole.

- It is recommended that COFUPRO generate an internal dialogue on the sources of the legitimacy of the PFs among farmers and civil society, beyond their access to state and federal authorities and to research institution administrators.

As they consolidated, the PFs increased their representation by including in their boards representatives of farmers' organizations and of some state-level system-product councils (see Section 6.4). Although this action has certainly made the boards more representative, it is not clear how representative they really are, given that their members are appointed by the boards themselves according to rules that are not always explicit. Neither is it clear how representative the product-systems are.

It must be recognized that representation is in no way essential for the foundations; it is enough that they be seen as legitimate (see above).[1] There are various measures the PFs can implement to bolster their legitimacy: more transparent selection of new council members, stronger collaboration with other agents in the national innovation system (beyond the traditional agricultural stakeholders), stronger contribution to the design and implementation of policies and programs to support innovation, and increased involvement in the transformation of the research system.

In the face of a crisis, the legitimacy that the academic community, together with new actors in and outside the agricultural sector could bestow, may well be an instrument key to the survival of the PFs, in the same way that the voices that defended INIFAP and the Colegio de Posgraduados in 2004 included farmers, other agriculture sector agents, and the scientific and technological community as a whole.[2]

- It is recommended that mechanisms to explore new instruments to foster agricultural innovation be incorporated into the operating rules of the PFs.

---

[1]For example, the boards of the international agricultural research centers are not representative, but they are legitimate as far as donors, governments, and research institutions are concerned. This legitimacy is based on the recognition gained by each council member.

[2]The 2004 federal budget proposal sent to the congress did not include resources for a number of public offices, including two major agricultural research organizations. Agricultural stakeholders strongly lobbied the congress and monies were allocated.

The PFs have tested new mechanisms to foster research and extension. These trials often contradicted the operational rules defined by SAGARPA. Such conflicts will reappear as long as the PFs continue exploring. The problem is not the exploration itself but the inflexibility of the rules. As mentioned in Section 2.1.2 (in the subsection "Defining Strategies in Complex Systems"), subjecting agents operating in evolving, ill-known complex processes to inflexible rules is a major impediment to innovation, because the latter do not allow decisionmakers to take advantage of emerging opportunities. One way of solving this contradiction is to allow the use of a fraction of the resources of the PFs to explore alternative innovation support mechanisms, even through noncompetitive mechanisms.

C.1.2 Defining the Target Group of the PFs

- It is recommended that a debate be promoted within the PFs, with other key actors in the agricultural innovation system (including farmer organizations and research and academic institutions renowned in the areas of rural development and innovation policy), and with the federal government about which farmers the PFs are currently targeting and whether this target is appropriate.

The PF founders thought that these organizations would represent farmers. However, there was no clear discussion of which farmers would be represented. The result, as happens in most countries, is that the PFs ended up representing and supporting mainly commercial farmers, because the poor farmers have great difficulty in articulating their needs. The reasons for these difficulties are (1) lack of human and social capital, especially the weakness of apolitical producers' associations and (2) the diversity of their needs, most of which are nontechnological in nature.

Some stakeholders in the agricultural sector have criticized the PFs for this orientation. It is not the clear, however, that the PFs should widen the focus of their actions. Support for small farmers, particularly poor ones, requires specific capabilities that the PFs do not have and can only acquire by investing heavily in additional staff and training. Furthermore, there is still no global consensus on what are the best means to support poor farmers. The difficulty of working with small farmers is evidenced by persistent worldwide poverty despite the billions of dollars spent annually on poverty-reduction programs and by the difficulty of expanding successful local experiences. In Mexico, the Ministry for Social Development, the state governments, and a number of NGOs are already working with poor rural households, and they are probably better trained for this than are the PFs.

If it is accepted that the foundations ought to support small farmers who are potential entrepreneurs, then the foundations will have to establish new alliances with NGOs and public-sector entities that work with these farmers. More crucially, however, they will have to develop the capacity to incorporate new methods of differentiating these particular farmers from those small farmers who do not have entrepreneurial capabilities. How to do this is still under debate in the international community, and no consensus has been reached yet. The literature on innovative capabilities in individuals (Renzulli 2003; Shavinina and Seeratan 2003) has shown that there are no exact methods for identifying people capable of innovating. In other words, identifying among all small farmers potential entrepreneurs will require methodologies that have not yet been developed.

Before expanding their activities in this direction, it is advisable to debate with a broad range of stakeholders the wisdom of such an expansion and the human and financial resources it would entail. Within the PFs, these activities would have to be the responsibility of a unit specialized in the analysis of new intervention mechanisms (see below).

C.1.3 Structure and Role of COFUPRO and the PFs

- It is recommended that a dialogue be established to redefine the relationship between COFUPRO and the PFs. In particular, it is recommended that COFUPRO have limited formal control over PF managers.

Relations between individual PFs and COFUPRO have been changing, with a gradual shift of power from the PFs to COFUPRO. However, there are no formal mechanisms by which COFUPRO can press individual PFs to avoid practices that are considered unacceptable by the PFs as a whole. This completely decentralized system reduces the effectiveness of COFUPRO, each PF, and the PFs as a set. Although it is not desirable that the PFs be totally dependent on COFUPRO, the present structure is not efficient, because it forces COFUPRO's president and executive secretary to invest too much time in lobbying (see Sections 5.5, 6.1.3, and 6.2.3). Currently the individual PFs accept the need for COFUPRO to coordinate and promote certain programs while the PFs themselves maintain a high degree of autonomy.

- It is recommended that the functions and duties of the president of each PF and of COFUPRO be explicitly defined. With this end in mind, it is helpful to be acquainted with the governance structures of other foundations that also finance research and extension activities (for example, the Bill and Melinda Gates Foundation or the Rockefeller Foundation).

As mentioned in Sections 5.5, 6.1.2, 6.2, and 6.3, the main task of the PF presidents is to interact with the state's political powers. Some presidents, though, feel that they should in addition manage the PF as if it were their own business. In most cases, the presidents feel that final responsibility for the functioning of the PF lies with them. In practice, however, the lines of action for the set of PFs are defined in the COFUPRO assembly and the interaction of the managers with COFUPRO. This confusion has various causes, including the presidential responsibility for the PF's assets, the president's power over the manager, and the organizational culture.

The biggest research-financing international foundations (for example, the Bill and Melinda Gates Foundation, the Rockefeller Foundation, and the Ford Foundation) have more differentiated management structures in which the board and its president are only responsible for defining the overall operational strategy, supervising a professional management team, and hiring the upper administrative echelons. The professionals are selected not only for their administrative capacity but also for their experience in specific fields of research (for example, natural resource management or plant breeding). With this background, managers and professionals are able to evaluate the quality and relevance of the project proposals they review.

- It is recommended that the abilities that PF presidents and managers should possess be explicitly defined.

As explained in Sections 6.1.2, 6.2, and 6.3, PF presidents and managers are expected to fulfill various requirements, some of which are implicit. For the presidents, the most important requirements are that they possess significant social capital, have good access to the state governor, and use these contacts for the PF's benefit. For the managers, the requirements are that they be loyal to the PF presidents, fulfill their obligations to other managers, support COFUPRO, be able to interact with local research institutions, and be good managers. When there is a conflict between loyalty to the president and professional capacity, it is generally resolved in favor of loyalty. Besides these qualifications, though, there are other abilities that PF authorities should possess, for instance, a basic acquaintance with management of research and extension.

Another problem with the capabilities managers should develop is that these are defined in terms of the individual PF's needs. Nevertheless, the managers also play an important role in COFUPRO and the set of PFs. As was mentioned above, selecting managers so that, as a group, they cover a wide range of capabilities would contribute to the benefit of all PFs. But this criteria would imply a further reduction in the control PF presidents have over their managers. This problem is essentially one of collective action: each PF needs to sacrifice some of its own independence to benefit from joint actions. The success of these actions is a function of the incentive structure to induce participation, the ability of the set to enforce agreements, and the costs and benefits each PF incurs by participating.

Lack of a clear presidential job description complicates selection, because each assembly member places a different value on the implicit and explicit job requirements. Similarly, the lack of clarity regarding desired qualifications for the manager and his or her subordinate hierarchy complicates professionalization of these individuals.

- It is recommended that PF managers be selected by a committee consisting of the PF president, the treasurer, and a representative of COFUPRO. Furthermore, managers should be hired for pre-established, renewable periods longer than the president's term.

The absolute formal dependence of the managers on the presidents, together with a partial functional dependence on COFUPRO, creates various disfunctionalities, among which stand out the difficulties of training managers and of inculcating best practices. The weakening of the link between president and manager and the hiring of the latter for a period longer than the presidential term would contribute to a more professional management. At the same time, maintaining the dependence of the manager on the PF (not necessarily on the president) will allow both the maintenance of links to state agriculture and the preservation of the decentralized structure that has contributed to the success of the PFs.

- It is recommended that a professional managerial career be created, which would be an incentive for the retention of staff trained by the PFs.

Learning in the PFs has been mainly concentrated on the managers, with the exception of a few presidents. But the exclusive dependence of managers on the presidents complicates the learning process for three reasons. First, because some presidents place a higher value on loyalty than on professional skills, when they take office, they change the manager. Thus the learning and experience of the departing manager are lost. Second, the incoming manager sometimes lacks the professional profile needed to be productively incorporated into the set of managers. Third, a significant portion of the learning process takes place in the course of interactions with other managers and with COFUPRO, but the managers at times have insufficient commitment to actions taken outside their own PFs. The creation of a professional career in which contributions to the set of PFs are recognized would encourage managers to remain in the job while fostering collective action.

The professional career should have two components: an adequate set of incentives and a training program. The incentive set should reward activities valued by the PFs. In practice, an incentive system is an agreement between the manager and the PF (or the set of PFs) that explicitly states the targets for the managers' increased efficiency and the PF's commitments to help the manager attain these targets. The targets should include criteria to measure the manager's actions and professionalism and specify the bonus or other compensation to reward the manager's contributions to the PF's success in attaining its goals.

The elaboration of a professional development program for managers should have two stages. First, the manager's job profile must be analyzed to define what capabilities are needed. Second, it is necessary to identify the current competencies of each of the existing managers. The development program should be elaborated on the bases of the discrepancies between what the managers need and what they actually have. The program should be carried out in coordination with an institution of higher education so that, should the manager so desire, an academic degree can be obtained (see below).

Training needs can be defined using a qualification matrix, a synoptic representation of the qualifications effectively available in management teams. The first step is the definition of the abilities a manager should have; this definition can be arrived at in a workshop involving the whole set or a subset of managers and outside facilitators. Each column of the matrix identifies a qualification, and the managers' names are placed in the rows. The team in charge of defining the training program marks for each manager his or her mastering of each of the qualifications. The matrix columns show the weaknesses or strengths of the management group in one of the competencies; the matrix rows indicate a manager's degree of versatility and mastery of his or her function. The columns show training needs for the managers as a group and the rows, individual manager needs. With the matrix data, a comprehensive development plan can be formulated, specifically adapted to each person and to the evolving needs of COFUPRO.

C.2 Learning Mechanisms

From their inception, some PFs have had a critical attitude toward the traditional research and extension approaches, the mechanisms the PFs themselves used to define priorities, and the routines developed to manage resources. This critical attitude led them to develop learning mechanisms in two key areas: improvement of administrative mechanisms and definition of research priorities. There is a third critical area, though, in which the PFs have not developed effective learning tools: new instruments to support research, extension, and innovation.

Among the contributions of the PFs to agricultural policy formulation, the exercise to prioritize agricultural chains and identify research demands stands out. The data generated by this activity had a great impact, because it was the only participative planning exercise for the agricultural sector that covered the whole country. Given the lack of other sources of information, federal and state governments used these priorities as the basis for their policies. In addition to generating information, the process was important because it opened channels of dialogue among actors who had previously been dissociated.

It will, however, be difficult for the PFs in the short run to generate other data of such impact. Communication channels among the actors are now well established, and as a result several agents have modified their patterns of interaction with others. When the identification of research demands was completed, it was judged to have been too expensive and complex to repeat, and an alternate data-updating mechanism was created: innovation units in each productive chain. These units used a smaller database and a methodology that was less participative than the original.

Two additional problems with the product-centered planning mechanism are that it reduces the flexibility of the PFs and hampers supporting emerging products. A parameter that SAGARPA has used to evaluate the foundations is the alignment of funded projects with identified demands. However, emergent products do not figure among the priorities, because by their very nature they are not yet important. In other words, the evaluation mechanism based on identified demands penalizes the more innovative PFs.

An area in which the PFs can make a significant contribution is in the design and implementation of new mechanisms to foster innovation, including support for the transformation of research institutions and the creation of effective extension mechanisms. Several PFs have implemented different extension mechanisms. These experiences are an exploration mechanism that has not yet been sufficiently exploited. In contrast, the PFs have not progressed with new ways of organizing research and innovation. In this and the next sections, we explore alternative ways of strengthening research and extension capabilities. Making a major contribution to the design of rural innovation programs will help to consolidate the legitimacy of the PFs in the agricultural sector, over and above the recognition they receive from the authorities. To make this contribution, though, the PFs will have to strengthen their analytical skills.

The two essential elements of the learning mechanisms developed by some of the PFs and COFUPRO were (1) the contributions of a few highly creative individuals and (2) the capacity for identifying knowledge gaps and finding external help to fill them. It should be stressed that the learning effort was very uneven; some PFs have experimented actively, whereas others have maintained a passive attitude.

Since its creation, COFUPRO has played a fundamental role in the development of learning mechanisms for the set of PFs, especially training, interfoundation exchange visits, interactions of a group of presidents and managers, exercise of self-evaluation, and contact with outsiders who brought key knowledge of research administration (especially Willem Janssen of ISNAR and José Laborde, a SAGARPA adviser).

The main instruments COFUPRO used to support learning by the PFs were the strategic planning exercises and the research-demand identification exercises. COFUPRO also played a fundamental part in some activities common to all PFs, such as the development of the research-demand identification methodology and the development of the online RFP.

- It is recommended that formal training mechanisms for the managers be established, for example, a graduate degree–granting academic program in the administration and financing of research, extension, and innovation.

The managers' capabilities are very uneven. The lack of formal, structured training mechanisms does not permit either the consolidation of a minimum ability base or the strengthening of the PF's culture. External certification of managers could help to rectify the present deficiencies in administrative capabilities. This certification could be structured around a core of short courses that includes such subjects as business administration, administration of STI, and researcher–farmer interaction techniques. The courses could be structured in such a way that those managers who so wished could obtain a master's degree, which would require forming an alliance with an institution of higher education able to award academic degrees.

To reduce per-student costs and widen the target group, the courses could be organized by COFUPRO and jointly financed with CONACYT. By so doing the program would also meet the needs of the Sectoral and Mixed Funds and the state science and technology councils. An additional advantage of these alliances is that the participation of students from different institutions helps to create horizontal links, increasing the efficiency of the system and the legitimacy of the PFs.

Competitively awarded funding to support research and innovation is to be found in nearly all Latin American countries. The master's program could also be opened to students from other Spanish-speaking countries.

- It is recommended that informal training mechanisms be created for PF managers and employees.

Formal training programs cannot transmit the tacit knowledge that managers acquire on the job. Thus an informal training program should also be established that could include extended training visits to PFs using best practices and distance learning courses (using, for example, the Internet).

- It is recommended that the training of presidents and managers be strengthened in the analysis, design, and implementation of research, extension, and innovation programs.

The PFs have developed a culture in which presidents do not need to know about scientific and technology policy, because these matters can be delegated to the managers (see Sections 6.1 and 6.2). This premise weakens COFUPRO's potential contribution to the design of research, extension, and innovation policies, because COFUPRO's interaction with the political hierarchy is centered on the presidents. The managers generally interact with the technical echelons.

Managers also do not receive any formal training in these areas; the acquisition of this type of knowledge is left up to the curiosity of each individual. However, as explained in Section 6.4, the area in which the PFs could make their greatest contribution is in the design and implementation of research and innovation programs. Training managers in these topics could take place within the framework of the above-recommended diploma courses.

Training the presidents in these subjects could take place in the form of a short course at the start of the presidential term followed by subsequent training sessions at COFUPRO events. At the moment, COFUPRO and some foundations already organize isolated training events; their impact would be enhanced by a more structured and organic character.

- It is recommended that a think-tank be created for the analysis of new mechanisms to foster research, extension, and innovation. The think-tank would serve to systematize and strengthen the learning process for the set of PFs and would also enable balancing immediate and long-term needs.

Since their beginning, the PFs have explored alternatives to improve their activities. On various occasions, exploration was based on external support that brought specialized knowledge that the PFs did not have (for example, from ISNAR and the Mexican Foundation for Total Quality). But institutional capabilities in these areas were not developed; rather, the exploration was concentrated in certain key individuals and in a few foundations.

COFUPRO's learning was focused on the consolidation of administrative procedures for the RFPs. This strategy is reaching its limits, because progress in the consolidation of a structure reduces creativity (see Section 2.3) and the challenges and opportunities that the PFs face today are more complex than those they initially faced. Limitations that the organizational structures impose on individual creativity have been recognized in the management literature, which recommends the creation of a specialized structure outside the routines of the mother organization (Christensen and Raynor 2003). This solution is also recommended for the PFs: create a formal think-tank to analyze scientific and technological policies and explore new programs to support research and innovation.

Another reason that learning mechanisms need to be strengthened is that several PFs recognize the limitations of the linear vision of science and have begun to look for new ways to foster innovation. There is, however, no worldwide consensus on what the most effective programs to support agricultural innovation are. In other words, the PFs are pushing up against the international frontier of knowledge in these areas. To continue using this conceptual framework, the PFs will have to develop their own analytical capabilities, because they will not be able to rely on others, as they did in the past.

Finally, the PFs and the managers developed solid capabilities for administering RFPs and projects. In contrast, most presidents became specialized in making contacts at the political level and with other presidents. Thus the foundations lack spaces specialized in the exploration of new instruments to support innovation.

The creation of a think-tank should be handled with care. The dependence of the PFs on key individuals, together with an ill-defined organizational structure, leaves them vulnerable to the departure of these key individuals. However, consolidated structures can reduce the creativity of individual PFs. As mentioned in Chapter 2 and Section 6.3, success depends on maintaining a balance between exploration and exploitation through the manipulation of variation (independent testing by the PFs) with an effective selection system (central analysis of individual experiences).

This think-tank should have a minimum structure (one or two trained professionals and an innovative farmer) and a budget that is small but allows the team to operate. The farmer would be responsible for guaranteeing smooth communications with the set of PFs and with other agents in the innovation system; the professionals would bring specialized knowledge and contacts with outside experts. To ensure close interactions with existing structures, the space could be a technical committee reporting to the president of COFUPRO. This committee would identify and systematize the innovations from each PF, organize monitoring of the experiences, help to diffuse the most effective innovations among the PFs, and propose innovations for the PFs to test. A small budget would force the professionals to negotiate projects with the PFs and with other actors in the innovation system.

- It is recommended that external experts be consulted to define a knowledge-administration strategy (in the broad sense of the term) for the set of PFs.

Since their establishment, the PFs have made important efforts to build databases (especially of technologies and projects). These efforts have been characterized by the PFs as "knowledge management." But this effort is too restrictive an interpretation of the concept.

The specialized literature has defined knowledge management as the set of decisions and systematic actions relating to all aspects of individual and collective knowledge. Knowledge management is perceived as a system that balances four knowledge dimensions: content, process, culture, and infrastructure (see Section 2.3.3). Content refers to the knowledge that can be managed. Processes include identification of needs, retrieval and utilization, creation or acquisition, collection, documentation and storage, and transference or socialization of knowledge. Culture refers to the organization's practices, form, frequency of activities, and attitudes toward those activities that affect the nature of the learning process at the individual, group, and organizational levels. Infrastructure refers to the establishment of a technical platform to share knowledge; the platform should include a material dimension (such as hardware and software) and a human dimension (such as the personnel to support the system's use and application of the knowledge-administration processes). Restricting knowledge management to databases amounts to reducing the system to infrastructure.

- It is recommended that a culture (and its attendant capabilities) be developed to document the activities that the PFs are exploring.

Both COFUPRO and the PFs document only a fraction of the actions and processes they implement. For example, some foundations are testing different mechanisms to encourage innovation, but these experiences are not systematically documented, analyzed, and socialized. Documenting experiences is important for three reasons. First, it facilitates institutional learning by codifying tacit knowledge. Second, it helps to disseminate the actions of the PFs and contributes to their legitimation in the eyes of researchers, farmers, and other actors in the innovation system. Third, writing about a process forces the writer to think about it, contributing to the consolidation of knowledge. Systematization of these experiences could be an activity shared between the above-proposed think-tank and PF managers.

C.3 Relationship with Research and Extension Institutions

- It is recommended that COFUPRO promote a dialogue on the future of the agricultural research system. This dialogue should include, apart from other actors, SAGARPA, the secretary of the treasury, and CONACYT.

The PFs are presently operating in a system in which some research institutions are being profoundly transformed, others are changing more gradually, and the extension mechanisms are still unarticulated. A major obstacle in transforming the research system is the difficulty in hiring new researchers. Although the solution to this problem is beyond the scope of the PFs, they can contribute by promoting a wide dialogue that explicitly includes authorities from the Ministry of the Treasury. The dialogue should be centered on the future of the agricultural research system, the human and financial resources necessary for the transformation and consolidation of this system, and mechanisms to bring about change. The debate could perhaps be organized by SNITT with support from the PFs, although at present SNITT does not have the structure to do it.

- It is recommended that pilot projects to explore alternative ways of contracting research and innovation activities be implemented to strengthen the best teams, facilitate the emergence of new high-quality research teams, and promote interdisciplinary projects.

Two types of support for research institutions should be tested. First, a portion of resources should be allocated to projects submitted by actors from the innovation system in a noncompetitive process. For these projects, a brief presentation will be needed, and then a group of managers should work with the presenters of the project to structure it in a way that would contribute to the objectives of the PFs. This mechanism could make the operation of the PFs more flexible, allowing identification of emerging products or processes that are not included in the prioritized chains (see Section 6.3). For this mechanism to be effective, individuals in the PFs who collaborate with researchers in project design must have a solid background in the disciplines in which they are operating. The Rockefeller and Ford Foundations use a similar method to select projects to finance. If the procedure proves to be successful, it could indicate new training areas for managers.

Second, the PFs could help to strengthen research capabilities through direct support of the strongest research teams and new mechanisms to foster the emergence of new teams of excellence. As explained in Sections 2.2 and 2.3, formal research contributes to innovation when (1) researchers establish close links with farmers (thereby assuring that the research is pertinent and the farmers can use the results) and (2) the research is of high quality (because otherwise it is not research). It is important, though, to recognize that quality is not necessarily correlated with publication in indexed journals but rather with the creation of original information that is useful for actors in the innovation system.

To strengthen existing teams, the PFs could offer the best researchers 5-year contracts, without the requirement of presenting research projects. The contract should permit supplementing salaries, require hiring young researchers, allow the employment of research assistants, and in some cases facilitate investments in equipment. The two key conditions of these contracts should be that (1) researchers work closely with innovative farmers (to ensure relevance) and (2) researchers' programs be periodically evaluated by a team that includes recognized foreign professionals (to ensure quality). The periodicity of the evaluations cannot be too short (less than 3 years), because the reviews are expensive, the researcher has to invest time in preparing materials for the evaluators, and it is necessary to allow sufficient time for the researcher to obtain results.

Similar contracts could be offered to emergent researchers, with the condition that they be affiliated to a recognized research institution. A more modest trajectory could be required of these researchers. While the first contract is in force (the consolidation period), quality requirements would be the same as for recognized researchers, but less productivity would be required, because these researchers need a grace period while they establish their lines of work.

These mechanisms should not be used to finance a graduate education, given that CONACYT has a scholarship program. To implement these procedures, it may be necessary to modify the operational rules of SITT, because awards would not be by open competition. Given that the effectiveness of these arrangements is not known, it is important to first experiment with pilot projects involving the think-tank and outside specialists, while the traditional RFPs are continued.

- It is recommended that more projects be supported in which recognized foreign researchers participate, as a means to create stable interactions between local and foreign researchers.

A complementary mechanism to strengthen research capabilities is by financing projects that include foreign researchers. The growing complexity of research and innovation activities forces individual actors to join international innovation networks (see Sections 2.2 and 2.3). Researchers who do not frequently interact with foreign colleagues from centers of excellence rapidly find themselves out of date.

The PFs are already supporting projects in which foreign researchers participate. It is recommended that these projects be further encouraged. Moreover, in priority areas in which there is no domestic capacity, first-rank foreign centers should be contracted, making sure that the contract stipulates development of local capacity.

- It is recommended that new mechanisms be explored to define priorities for the PFs as a whole.

COFUPRO coordinated the definition of research priorities in 2002 and 2003. After this process there were no further attempts to define priorities in a coordinated fashion, although COFUPRO's innovation units have updated information for various chains. At present each PF defines its priorities in its own way, sometimes annually. The result is that priorities vary across states, but more importantly, across time, hampering the consolidation of research programs.

Another problem with the methods used to define research priorities is that they do not always allow the participation of the researchers themselves. As mentioned in Section 2.2, innovation systems evolve through the interactions of supply and demand. A demand-oriented system is short-sighted and cannot identify emergent opportunities that only researchers know of.[3]

Priorities should be defined with a mechanism that is efficient in the sense that it is not excessively expensive, allows the participation of a broad range of innovative actors, and permits identification of emergent opportunities. To identify an adequate mechanism, it is possible to evaluate what individual PFs have done and what is used in other countries (particularly those that belong to the Organisation for Economic Co-operation and Development) and by public research institutes from developed countries (such as the Agricultural Research Service of the United States Department of Agriculture or Agrifood Canada). These institutions should be analyzed, because they are technological centers that support farmers rather than pure research institutions. Also recommended is the evaluation of new planning mechanisms for complex systems that have been proposed in the literature in the past decade (see Section 2.1.2).

- It is recommended that the state technical committees be replaced by regional or national technical committees that should also incorporate foreign researchers.

At present each PF has its own technical committee to evaluate proposals. Committee members are often chosen for their affiliation to a particular institution and not for their technical capacity. Given the small size of the state research systems, it is difficult for each state to establish a highly qualified technical committee. The creation of regional or national committees by broad subject areas could contribute to solving this problem.[4] To create regional committees, it is possible to use the committees COFUPRO has already established. Foreign researchers should be invited to participate, thus increasing transparency in the evaluations.

- It is recommended that the PFs develop capabilities for the analysis of innovation mechanisms, especially to foster the emergence and consolidation of networks and the development of entrepreneurial skills in small farmers. It is also recommended that a system of experimentation coordinated by COPUFRO be established to test new extension methods.

Currently Mexico does not have agricultural extension programs of recognized effectiveness in all states and productive chains. The reason is that different types of farmers have different needs and capabilities. The diversity of the PFs and their local roots could be an important factor in identifying adequate extension mechanisms for specific groups of farmers and specific conditions. To identify these mechanisms, the PFs need to professionalize their managers so that they can develop capabilities in the analysis of research and extension programs. In addition, COFUPRO must play the role of facilitator in the diffusion of experiences among the PFs.

The literature on innovation suggests that extension mechanisms should not be restricted to technical topics; they should instead emphasize the facilitation of collective action and the integration of innovation networks (see Sections 2.2 and 2.3). At present, there are no recognized procedures for promoting the emergence and consolidation of innovation networks in agriculture, especially networks that include small farmers. Some PFs have

[3]Researchers know best the potential of their own lines of research. Their interactions with other actors of the innovation system can convert this potential to useful innovations when used by social or productive agents. Several developed countries define research priorities every 5 years through Delphi surveys that include only researchers. In the more than 40 years that the system has been used, approximately 85 percent of the priorities turned out to be correct.

[4]For example, national technical committees for specific topics (such as crops, cattle rearing, or forestry) can be created.

experimented with different ways of promoting these networks, but there is no systematic, coordinated effort.

- It is recommended that COFUPRO promote a debate on the SAGARPA-CONACYT Sectoral Fund, in particular its vision of science and mode of operation.

The RFP structure of the SAGARPA-CONACYT Sectoral Fund and the information required in the application documents reflect a simplistic and mechanistic vision of research. For example, it is necessary to provide the titles of articles to be published before starting the research. If researchers are really able to provide this information, the research is on the point of being concluded or is not really contributing anything new and therefore is not research. If they intend to conduct groundbreaking research, researchers simply supply spurious documentation for these article titles, so the information is not useful. Another example of this mechanistic vision is the requirement to specify the expected quantitative, qualitative, scientific, economic, social, and environmental impacts. As explained in Sections 2.1 and 2.3, research processes are by nature uncertain. Effects can only be predicted with a degree of certainty when projects are either not new or merely incremental. In effect, selection criteria that are based on expected outcomes only conspire against the most innovative research. In the case of innovative projects, the ex ante effects can only be specified in very broad terms, which is why the information is not useful for decisionmaking. A last example is the excessive detail that is requested in the budget justification, whereas the analysis of mechanisms of interaction among researchers and other actors in the innovation system is neglected.

In sum, project selection is in part based on information that is irrelevant to the real evaluation of the research potential. In contrast, information that is truly relevant (mechanisms of interaction with other actors in the innovation system, research quality, or information that would permit the identification of conflicts of interest with the reviewers) is either not requested or requested in insufficient detail. An additional problem with this mechanism is that it forces researchers to waste a great deal of time generating information that is irrelevant. A study should be made of mechanisms used by the administrators of other funds (competitive or noncompetitive, domestic and abroad) to address these problems.

APPENDIX **D**

Topics Covered in the Interviews

D.1 History of the PFs and COFUPRO

- The process that led to the creation of the PFs and identification of different stages
- Expectations of decisionmakers regarding the role the PFs would play in the agricultural research system
- Identification of the initial dynamics of the PFs
- The process that led to the creation of COFUPRO and its subsequent evolution
- Reactions of the different actors in the innovation and research systems to the creation of the PFs
- The evolving relations between the PFs and COFUPRO on the one hand, and the federal and state governments on the other
- Identification of the main problems faced by the PFs and COFUPRO and of the strategies they followed to solve them

D.2 Organizational Culture and Governance

- Main assumptions in the emergence of an organizational culture
- The current roles of SAGARPA, farmers (including presidents and managers), and state governments in the governing bodies of COFUPRO and the PFs
- Implicit and explicit rules for the designation of COFUPRO's presidents and executive secretaries
- Implicit and explicit rules for the designation of PF presidents and managers
- Evolution of the procedures for priority setting and resource allocation in the PFs and COFUPRO
- Identification of leadership styles in COFUPRO, in particular, analysis of the power-sharing arrangements between the president and the executive secretary
- Identification of the conflicts that exist or existed in COFUPRO or between COFUPRO and the PFs, and of the strategies to solve them
- Financing mechanisms

D.3 Learning Activities

- Evolution of the RFPs and of the interaction with research institutions
- Evolution of priority-setting methods
- Evolution of internal and external methods to evaluate COFUPRO and the PFs
- Creation of a language common to all PFs
- Identification of different instruments tried by COFUPRO and the PFs
- Strategies to document positive experiences and to learn from them
- Training of authorities, employees, and farmers in COFUPRO and the PFs
- Control of projects financed by COFUPRO and the PFs
- Strategies to learn about the dynamics of innovation systems and identify instruments to foster innovation
- Strategies to learn about STI policies (including extension)

D.4 Impact on the Research and Innovation Systems

- Evolution of the interactions between COFUPRO and the PFs on the one side, and research institutions, policymakers, and researchers on the other
- Evolution of the interaction between COFUPRO and CONACYT, especially regarding priority setting and resource allocation
- Influence of COFUPRO and the PFs on research institutions
- Influence of COFUPRO and the PFs on individual researchers, especially on their working routines
- Influence of COFUPRO and the PFs on extension programs

References

Adams, R. N. 1993. Energía y estructura. Una teoría del poder social. Mexico City: Fondo de Cultura Económica.

Alex, G., W. Zijp, and D. Byerlee, with the collaborations of the AKIS thematic group. 2002. *Rural extension and advisory services: New directions.* Rural Development Strategy Background Paper 9. Washington, D.C.: World Bank.

Alston, J. M., G. W. Norton, and P. G. Pardey. 1995. *Science under scarcity: Principles and practice of agricultural research evaluation and priority setting.* Ithaca, N.Y., U.S.A.: Cornell University Press.

Appelyard, M. M., N. W. Hatch, and D. Mowery. 2000. Managing the development and transfer of process technologies in the semiconductor manufacturing industry. In *The nature and dynamics of organizational capabilities,* ed. G. Dosi, R. R. Nelson, and S. G. Winter. New York: Oxford University Press.

Archibugi, D., and J. Michie. 1998. Trade, growth and technical change: What are the issues? In *Innovation policy in a global economy,* ed. Archibugi, D., J. Howells, and J. Michie. Cambridge: Cambridge University Press.

Archibugi, D., J. Howells, and J. Michie. 1999. Innovation systems and policy in a global economy. In *Innovation policy in a global economy,* ed. D. Archibugi, J. Howells, and J. Michie. Cambridge: Cambridge University Press.

Argote, L., and E. Darr. 2000. Repositories of knowledge in franchise organizations. In *The nature and dynamics of organizational capabilities,* ed. G. Dosi, R. R. Nelson, and S. G. Winter. New York: Oxford University Press.

Arnold, E., M. Bell, J. Bessant, and P. Brimble. 2000. Enhancing policy and institutional support for industrial technology development in Thailand—The overall policy framework and the development of the industrial innovation system. Paper prepared for the National Science and Technology Development Board, Thailand, with support from the World Bank. http://siteresources.worldbank.org/INTTHAILAND/Resources/333200-1163047690503/3140686-1163047766322/Arnold_2006_enhancing_policy_and_institutional_support_for_industrial_technology_development_in_thailand_vol.1.pdf. http://www.technopolis.co.uk/reports. Accessed June 10, 2007.

Axelrod, R., and M. D. Cohen. 1999. *Harnessing complexity. Organizational implications of a scientific frontier.* New York: Free Press.

Bailey, J. R., and C. M. Ford. 2003. Innovation and evolution: Managing tensions within and between the domains of theory and practice. In *The international handbook on innovation,* ed. L. V. Shavinina. London: Pergamon.

Bak, P. 1996. *How Nature works. The science of self-organized criticality.* New York: Springer-Verlag.

Barrett, C., T. Reardon, and P. Webb. 2001. Nonfarm income diversification and household livelihood strategies in rural Africa: Concepts, dynamics and policy implications. *Food Policy* 26: 315–331.

Bellon, M. R. 2001. *Participatory research methods for technology evaluation: A manual for scientists working with farmers.* Mexico City: Centro Internacional de Mejoramiento de Maíz y Trigo.

Berdegué, J. A. 2005. *Pro-poor innovation systems.* Background paper. Rome: International Fund for Agricultural Development.

Biggs, S. 2005. Learning from the positive to reduce rural poverty: Institutional innovations in agricultural and natural resources research and development. Paper prepared for the Impact Assessment Workshop, Mexico City, October 19–21, 2005.

Biggs, S., and H. Matsaert. 2004. *Strengthening poverty reduction programmes using an actor-oriented approach: Examples from natural resources innovation systems.* AgRen Network Paper 134. London: AgRen.

Bobbio, N., and N. Matteucci. 1988. *Diccionario de política.* Mexico City: Siglo XXI.

Branscom, L. M. 1999. From science policy to research policy. In *Investing in innovation: Creating research and innovation policy that works,* ed. L. M. Branscomb and J. H. Keller. Cambridge, Mass., U.S.A.: MIT Press.

Branscom, L. M., and R. Florida. 1999. Challenges to technology policy in a changing world economy. In *Investing in innovation: Creating research and innovation policy that works,* ed. L. M. Branscomb and J. H. Keller. Cambridge, Mass., U.S.A.: MIT Press.

Buchanan, M. 2002. *Nexus. Small worlds and the groundbreaking theory of networks.* New York: Norton.

Bush, V. 1945. *Science—The endless frontier: A report to the president on a program for postwar scientific research.* Washington, D.C.: Office of Scientific Research and Development.

Byerlee, D., G. Alex, and R. G. Echeverría. 2002. The evolution of public research systems in developing countries: Facing new challenges. In *Agricultural research policy in an era of privatization,* ed. D. Byerlee and R. G. Echeverría. Wallingford, U.K.: CAB International.

Calvin, L., and V. Barrios. 1999. Marketing winter vegetables from Mexico. *Journal of Food Distribution Research* 30 (1): 50–62.

Carayannis, E. G., E. Gonzalez, and J. J. Wetter. 2003. The nature and dynamics of discontinuous and disruptive innovations from a learning and knowledge management perspective. In *The international handbook on innovation,* ed. L. V. Shavinina. London: Pergamon.

Carroll, L. 1982. *The complete illustrated works of Lewis Carroll.* New York: Avenel.

Chambers, R., A. Pacey, and L. A. Thrupp, eds. 1989. *Farmers first: Farmer innovation and agricultural research.* Warwickshire, U.K.: Practical Action.

Charles, D. 2001. *Lords of the harvest: Biotechnology, big money, and the future of food.* Cambridge, Mass., U.S.A.: Perseus.

Christensen, C. M. 2003. *The innovator's dilemma: The revolutionary national bestseller that changed the way we do business.* New York: Harper Business Essentials, HarperCollins.

Christensen, C. M., and M. E. Raynor. 2003. *The innovator's solution. Creating and sustaining successful growth.* Cambridge, Mass., U.S.A.: Harvard Business School Press.

Christensen, C. M., S. D. Anthony, and E. A. Roth. 2004. *Seeing what's next. Using the theories of innovation to predict industry change.* Cambridge, Mass., U.S.A.: Harvard Business School Press.

COFUPRO (Coordinadora Nacional de Fundaciones Produce). 2001. Propuesta de modernización y cambio institucional 2001–2006. Working paper, Mexico City.

Cohen, W. M., and D. A. Levinthal. 1990. Absorptive capacity: A new perspective on learning and innovation. *Administrative Science Quarterly* 35: 128–152.

Consejo Estatal de Ciencia y Tecnología de Puebla and Fundación Produce Puebla. 2004. *Cadenas agroalimentarias. El papel estratégico de la tecnología y su prospectiva en el estado de Puebla.* Mexico City.

Coriat, B. 2000. The "abominable Ohno Production System." Competences, monitoring, and routines in Japanese production systems. In *The nature and dynamics of organizational capabilities*, ed. G. Dosi, R. R. Nelson, and S. G. Winter. New York: Oxford University Press.

Crozier, M. 1994. *L'entreprise à l'écoute. Apprendre le management post-industriel.* Paris: Editions du Seuil.

Crozier, M., and E. Friedberg. 1977. *L'acteure et le système.* Paris: Editions du Seuil.

Crutchfield, J. P., and P. Schuster, eds. 2003. *Evolutionary dynamics. Exploring the interplay of selection, accident, neutrality and function.* New York: Oxford University Press.

Davenport, T., and L. Prusak. 1998. *Working knowledge: How organizations manage what they know.* Cambridge, Mass., U.S.A.: Harvard Business School Press.

Davila, T., M. J. Epstein, and R. Shelton. 2006. *Making innovation work. How to manage it, measure it, and profit from it.* Upper Saddle River, N.J., U.S.A.: Wharton School Publishers.

Davis, B., A. de Janvry, E. Sadoulet, and T. Diehl. 2000. Poverty in the Ejido sector. *Investigación Económica* 232: 123–154.

Dixit, A. K., and R. S. Pindyck. 1994. *Investment under uncertainty.* Princeton, N.J., U.S.A.: Princeton University Press.

Dodgson, M. 1993. Organizational learning: A review of some literatures. *Organizational Studies* 14 (3): 375–394.

Dosi, G., B. Coriat, and K. Pavitt. 2000. Competences, capabilities and corporate performances. Dynacom Working Papers Series. Pisa: Sant'Anna School of Advanced Studies, Laboratory of Economics and Management.

Dosi, G., P. Llerena, and M. Sylos Labini. 2006. The relationships between science, technologies and their industrial exploitation: An illustration through the myths and realities of the so-called "European paradox." *Research Policy* 33: 1450–1464.

Dosi, G., R. R. Nelson, and S. G. Winter. 2000. Introduction: The nature and dynamics of organizational capabilities. In *The nature and dynamics of organizational capabilities,* ed. G. Dosi, R. R. Nelson, and S. G. Winter. New York: Oxford University Press.

Dutrénit, G. 2001. El papel de las rutinas en la codificación del conocimiento en la firma. *Análisis Económico* 16 (34): 211–230.

———. 2002. Debilidades en los procesos de la administración del conocimiento. El caso de una firma mexicana. *Investigación Económica* 241 (July–Sept.): 35–79.

EC (European Community). 2003. *Innovation tomorrow—Raising EU R&D intensity, improving the effectiveness of public support mechanisms for private sector research and development, direct measures.* EUR 20716. Brussels.

Echeverría, R. G. 1998. Will competitive funding improve the performance of agricultural research? Working Paper 98-16. The Hague: International Service for National Agricultural Research.

Echeverría, R. G., and H. Elliot. 2002. Financing agricultural research by competitive funds. In *Agricultural research policy in an era of privatization,* ed. D. Byerlee and R. G. Echeverría. London: CAB International.

Ekboir, J. M. 2002. Developing no-till packages for small-scale farmers. In *World wheat overview and outlook,* ed. J. M. Ekboir. Mexico City: Centro Internacional de Mejoramiento de Maíz y Trigo.

———. 2003. Why impact analysis should not be used for research evaluation and what the alternatives are. *Agricultural Systems* 78 (2): 166–184.

———. 2004. *Informe de Evaluación Nacional. Subprograma de investigación y transferencia de tecnología.* Mexico City: Food and Agriculture Organization of the United Nations. http://www

.fao-evaluacion.org.mx/Eval2003/Resultados2003/pdf/Informes_Nacionales/NAL_ITT_2003.pdf. Accessed June 3, 2009.

Ekboir, J. M., K. Boa, and A. A. Dankyi. 2002. *Impact of no-till technologies in Ghana.* CIMMYT Economics Paper 02-01. Mexico City: Centro Internacional de Mejoramiento de Maíz y Trigo.

Ekboir, J. M., J. A. Espinoza García, J. J. Arellano Espinoza, G. Moctezuma Lopez, and A. Tapia Naranjo. 2003. Análisis del sistema de investigación agropecuario Mexicano (Analysis of the Mexican agricultural research system). CIMMYT Economics Program Working Paper 03-01. Mexico City: Centro Internacional de Mejoramiento de Maíz y Trigo.

Etkin, J. 2000. *Política, gobierno y gerencia de las organizaciones.* Buenos Aires, Argentina: Prentice-Hall.

Eun, J. H., K. Lee, and G. Wu. 2006. Explaining the "university-run enterprises" in China: A theoretical framework for university–industry relationships in developing countries and its application to China. *Research Policy* 35: 1329–1346.

Fagerberg, J. 2005. Innovation: A guide to the literature. In *The Oxford handbook of innovation,* ed. J. Fagerberg, D. C. Mowery, and R. R. Nelson. Oxford and New York: Oxford University Press.

Fagerberg, J., and M. M. Godinho. 2005. Innovation and catching-up. In *The Oxford handbook of innovation,* ed. J. Fagerberg, D. C. Mowery, and R. R. Nelson. Oxford and New York: Oxford University Press.

Ferris, S., E. Faganzi, R. Best, C. Ostertag, M. Lundy, and T. Wandschneider. 2006. *A market facilitator's guide to participatory agroenterprise development.* Enabling Rural Innovation (ERI) Guide 2. Cali, Colombia: Centro Internacional de Agricultura Tropical.

Foro Consultivo Científico y Tecnológico. 2006. *Diagnóstico de la política científica, tecnológica y de fomento a la innovación en México (2000–2006).* Mexico City.

Foucault, M. 1978. *El discurso del poder.* Madrid: Crítica.

Freud, S. 1963. *Dora: An analysis of a case of hysteria.* New York: Collier.

Gabre-Madhin, E. Z., and S. Haggblade. 2004. Successes in African agriculture: Results of an expert survey. *World Development* 32 (5): 745–766.

Gallagher, R., and T. Appenzeller. 1999. Introduction: Beyond reductionism. *Science* 284: 79.

García, E., and L. Sanz-Menéndez. 2005. Competition for funding as an indicator of research competitiveness. *Scientometrics* 64 (3): 271–300.

Georgsdottir, A. S., T. I. Lubart, and I. Get. 2003. The role of flexibility in innovation. In *The international handbook on innovation,* ed. L. V. Shavinina. London: Pergamon.

Gibson, J. L., J. M. Ivancevich, and J. H. Donnelly. 1996. *Las organizaciones: Comportamiento, estructura, procesos,* eighth edition. Madrid: Irwin.

Gill, G., and D. Carney. 1999. *Competitive agricultural technology funds in developing countries.* Natural Resource Perspectives 41. London: Overseas Development Institute.

Gittleman, M., and E. N. Wolff. 1998. R&D activity and cross-country growth comparisons. In *Trade, growth and technical change,* ed. G. Archibugi and D. Michie. Cambridge: Cambridge University Press.

Goldberg, P. K., and N. Pavcnik. 2007. Distributional effects of globalization in developing countries. *Journal of Economic Literature* 45 (March): 39–82.

Hall, A., N. Clark, and G. Naik. 2007. Technology supply chain or innovation capacity? Contrasting experiences of promoting small-scale irrigation technology in South Asia. Working Paper 2007-014. Maastricht: United Nations University–Maastricht Economic and Social Research and Training Center on Innovation and Technology.

Hall, A., G. Bockett, S. Taylor, M. V. K. Sivamohan, and N. Clark. 2001. Why research partnerships really matter: Innovation theory, institutional arrangements and implications for developing new technology for the poor. *World Development* 29 (5): 783–797.

Handy, C. 1995. *Gods of management: The changing work of organizations.* London: Arrow.

Hartwich, F., J. Tola, A. Engler, C. González, G. Ghezan, J. M. Vázquez-Alvarado, J. A. Silva, J. J. Espinoza, and M. V. Gottret. 2008. *Building public–private partnerships for agricultural innovation.* Food Security in Practice Technical Guide 4. Washington, D.C.: International Food Policy Research Institute.

Henderson, R., and I. Cockburn. 2000. Measuring competence? Exploring firm effects in drug discovery. In *The nature and dynamics of organizational capabilities,* ed. G. Dosi, R. R. Nelson, and S. G. Winter. New York: Oxford University Press.

Hirvonen, M. 2007. A tourist guide to systems studies of rural innovation. Learning, Innovation, Knowledge (LINK) Policy Resources on Rural Innovation 1. http://innovationstudies.org/images/stories/linktouristguide.pdf. Accessed May 4, 2009.

Holden, S., B. Shiferaw, and J. Pender. 2004. Non-farm income, household welfare, and sustainable land management in a less-favored area in the Ethiopian highlands. *Food Policy* 29: 369–392.

Huffman, W. E., and R. E. Evenson. 2003. New econometric evidence on agricultural total factor productivity determinants: Impact of funding sources. Working Paper 03029. Ames, Iowa, U.S.A.: Iowa State University, Department of Economics.

Huffman, W. E., and R. Just. 1999. Agricultural research: Benefits and beneficiaries of alternative funding mechanisms. *Review of Agricultural Economics* 21 (1): 2–19.

———. 2000. Setting efficient incentives for agricultural research: Lessons from principal-agent theory. *American Journal of Agricultural Economics* 82 (4): 828–841.

IDRC (International Development Research Center). 1997. A decade of reform: Science and technology policy in China. http://www.idrc.ca/en/ev-9360-201-1-DO_TOPIC.html. Accessed April 23, 2009.

IFAD (International Fund for Agricultural Development). 2008. Sending money home: Worldwide remittance flows to developing countries. http://www.ifad.org/events/remittances/maps/index.htm. Accessed February 25, 2008.

Kauffman, S. 1995. *At home in the universe. The search for the laws of self-organization and complexity.* New York: Oxford University Press.

———. 2000. *Investigations.* New York: Oxford University Press.

Kraemer, S. 2006. *Science and technology policy in the United States.* New Brunswick, N.J., U.S.A.: Rutgers University Press.

Laudel, G. 2005. Is external research funding a valid indicator for research performance? *Research Evaluation* 14 (1): 27–34.

Leonard-Barton, D. 1995. *Wellsprings of knowledge.* Cambridge, Mass., U.S.A.: Harvard Business School Press.

Lepori, B., P. van den Besselaar, M. Dinges, B. Potì, E. Reale, S. Slipersæter, J. Thèves, and B. van der Meulen. 2007. Comparing the evolution of national research policies: What patterns of change? *Science and Public Policy* 34 (6): 372–388.

Levinthal, D. 2000. Organizational capabilities in complex worlds. In *The nature and dynamics of organizational capabilities,* ed. G. Dosi, R. R. Nelson, and S. G. Winter. New York: Oxford University Press.

López-Córdova, E., and A. Olmedo. 2006. *International remittances and development: Existing evidence, policies and recommendations.* Occasional Papers 41. Washington, D.C.: Instituto para la Integración de América Latina y el Caribe.

Luhmann, N. 1997. *Organización y decisión. Autopoiesis, acción y entendimiento comunicativo.* Barcelona: Anthropos and Universidad Iberoamericana.

Lundvall, B. A. 1992. *National systems of innovation: Towards a theory of innovation and interactive learning.* London: Pinter.

Lundvall, B. A., and S. Borrás. 2005. Science, technology and innovation policy. In *The Oxford handbook of innovation,* ed. J. Fagerberg, D. C. Mowery, and R. R. Nelson. Oxford and New York: Oxford University Press.

Machamer, P. 2002. A brief historical introduction to the philosophy of science. In *The Blackwell guide to the philosophy of science,* ed. P. Machamer and M. Silberstein. Oxford: Blackwell.

Malerba, F. 1993. The national system of innovation: Italy. In *National innovation systems. A comparative analysis,* ed. R. Nelson. New York: Oxford University Press.

McClellan, J. E., III, and H. Dorn. 1999. *Science and technology in world history.* Baltimore, Md., U.S.A.: Johns Hopkins University Press.

Meinzen-Dick, R., M. Adato, L. Haddad, and P. Hazell. 2004. *Science and poverty. An interdisciplinary assessment of the impact of agricultural research.* Food Policy Report 16. Washington, D.C.: International Food Policy Research Institute.

Miller, A. I. 2000. *Insight of genius. Imagery and creativity in science and art,* first edition. Cambridge, Mass., U.S.A.: MIT Press.

Mintzberg, H. 1999. *Le management,* second edition. Paris: Éditons d'Organisation.

Mowery, D. C., and N. Rosenberg. 1979. The influence of market demand upon innovation: A critical review of some recent empirical studies. *Research Policy* 8 (2): 102–153.

Mowery, D. C., and B. N. Sampat. 2005. Universities in national innovations systems. In *The Oxford handbook of innovation,* ed. J. Fagerberg, D. C. Mowery, and R. R. Nelson. Oxford and New York: Oxford University Press.

Muñoz, M. 2005. *Informe de evaluación nacional. Subprograma de Investigación y Transferencia de Tecnología.* Rome: Food and Agriculture Organization of the United Nations.

Nelson, R. R., ed. 1993. *National innovation systems, a comparative analysis.* New York: Oxford University Press.

———. 1998. El Ambiente Cambiante para "Aprender de los Demás." Paper presented at the meeting on Diffusion, Assimilation and Use of Technology by Firms. Washington, D.C., February 9–10, 1998.

Nelson, R. R., and N. Rosenberg. 1993. Technical innovation and national systems. In *National innovation systems, a comparative analysis,* ed. R. R. Nelson. New York: Oxford University Press.

Nelson, R. R., and S. G. Winter. 1982. *An evolutionary theory of economic change.* Cambridge, Mass., U.S.A.: Harvard University Press.

Nickles, T. 2003. Evolutionary models of innovation and the Meno problem. In *The international handbook on innovation,* ed. L. V. Shavinina. London: Pergamon.

Nicolis, G., and I. Prigogine. 1989. *Exploring complexity.* New York: W. H. Freeman.

Nonaka, I., and H. Takeuchi. 1995. *The knowledge-creating company.* New York: Oxford University Press.

OECD (Organisation for Economic Co-operation and Development). 1996. *The knowledge-based economy.* OCDE/GD(96)102. Paris.

———. 1999. *Managing national innovation systems.* Paris.

Olson, E. E., and G. H. Eoyang. 2001. *Facilitating organization change. Lessons from complexity science.* San Francisco: Jossey-Bass/Pfeiffer.

Panebianco, A. 1990. *Modelos de partido.* Madrid: Alianza Universidad.

Pardey, P. G., J. M. Alston, and R. R. Piggott, eds. 2006. *Agricultural R&D in the developing world: Too little, too late?* Washington, D.C.: International Food Policy Research Institute.

Pavitt, K. 2005. Innovation processes. In *The Oxford handbook of innovation,* ed. J. Fagerberg, D. C. Mowery, and R. R. Nelson. Oxford and New York: Oxford University Press.

Pérez, R. 2003. *La gouvernance de l'entreprise.* Paris: Éditions la Découverte.

PF and COFUPRO (Fundaciones Produce and Coordinadora Nacional de Fundaciones Produce). 2001. Propuesta de modernización y Cambio Institucional 2001–2006. Working paper. Mexico City.

Polanco Jaime, A. 1996. Los retos institucionales de la innovación tecnológica. In *Posibilidades para el Desarrollo Tecnológico del Campo Mexicano,* ed. J. L. Solleiro, M. C. del Valle, and E. Moreno. Mexico City: Editorial Cambio XXI.

Powell, W. W., and S. Grodal. 2005. Networks of innovators. In *The Oxford handbook of innovation,* ed. J. Fagerberg, D. C. Mowery, and R. R. Nelson. Oxford and New York: Oxford University Press.

Ragin, C. C. 2000. *Fuzzy-set social science.* Chicago: University of Chicago Press.

Reardon, T. 2005. *Retail companies as integrators of value chains in developing countries: Diffusion, procurement system change, and trade and development effects.* Eschborn, Germany: Deutsche Gesellschaft für Technische Zusammenarbeit.

Reardon, T., and L. Flores. 2006. "Customized competitiveness" strategies for horticultural exporters: Central America focus with lessons from and for other regions. *Food Policy* 31 (6): 483–503.

Reij, C., and A. Waters-Bayer. 2001. An initial analysis of farmer innovators and their innovations. In *Farmer Innovation in Africa: A Source of Inspiration for Agricultural Developmen,* ed. C. Reij and A. Waters-Bayer. London: Earthscan.

Reis, S. M., and J. S. Renzulli. 2003. Developing high potentials for innovation in young people through the schoolwide enrichment model. In *The international handbook on innovation,* ed. L. V. Shavinina. London: Pergamon.

Renzulli, J. S. 2003. The three-ring conception of giftedness: Its implications for understanding the nature of innovation. In *The international handbook on innovation,* ed. L. V. Shavinina. London: Pergamon.

Richters, J. 1997. The Hubble hypothesis and the developmentalist's dilemma. *Development and Psychopathology* 9: 193–230.

Rivera, W. M., and G. Alex, eds. 2004. *Volume 1. Decentralized systems. Case studies of international initiatives.* Agriculture and Rural Development Discussion Paper 8, Extension Reform for Rural Development. Washington, D.C.: World Bank.

———. 2005. Extension reform: The challenges ahead. In Proceedings of the Association for International Agricultural and Extension Education Annual Meeting, San Antonio, Tex., U.S.A., May 25–31.

Robbins, S. 1999. *Comportamiento organizacional,* eighth edition. Mexico City: Prentice-Hall Pearson.

Robson, A. J. 2005. Complex evolutionary systems and the Red Queen. *Economic Journal* 115 (June): F211–F224.

Rodrik, D. 2006. Goodbye Washington Consensus. Hello Washington confusion? A review of the World Bank's *Economic growth in the 1990s: Learning from a decade of reform. Journal of Economic Literature* 44 (4): 973–987.

Root-Bernstein, R. 2003. Problem generation and innovation. In *The international handbook on innovation,* ed. L. V. Shavinina. London: Pergamon.

Rothwell, R. 1994. Industrial innovation: Success, strategy, trends. In *The handbook of industrial innovation,* ed. M. Dodgson and R. Rothwell. Cheltenham, U.K.: Edward Elgar.

Rubenstein, K. D., P. W. Heisey, C. Klotz-Ingram, and G. B Frisvold. 2000. Competitive grants and the funding of agricultural research in the U.S. Paper presented at the annual meeting of the American Association of Agricultural Economics, Tampa, Fla., U.S.A., July 30–August 2, 2000.

Rycroft, R. W., and D. E. Kash. 1999. *The complexity challenge: Technological innovation for the 21st century.* Science, Technology, and the International Political Economy Series A. New York: Cassell Imprint.

Savall, H., and V. Zardet. 1995. *L'ingénierie stratégique du Roseau.* Paris: Economica.

Schein, E. H. 1984. Coming to a new awareness of organizational culture. *Sloan Management Review* 25: 3–16.

———. 1991. *Organizational culture and leadership,* second edition. San Francisco and Oxford: Jossey-Bass.

———. 1996. *Strategic pragmatism. The culture of Singapore's Economic Development Board.* Cambridge, Mass., U.S.A.: MIT Press.

Shapira, P. 1999. Manufacturing extension: Performance, challenges and policy issues. In *Investing in innovation: Creating research and innovation policy that works,* ed. L. M. Branscomb and J. H. Keller. Cambridge Mass., U.S.A.: MIT Press.

Shavinina, L. V., and K. L. Seeratan. 2003. On the nature of individual innovation. In *The international handbook on innovation,* ed. L. V. Shavinina. London: Pergamon.

Simon, H. A. 1996. Bounded rationality and organizational learning. In *Organizational learning,* ed. M. D. Cohen and L. S. Sproull. Thousand Oaks, Calif., U.S.A.: Sage.

Smit, M. 2007. *We're too much in "to do" mode: Action research into supporting international NGOs to learn.* Praxis Paper 16. The Hague: International NGO Training and Research Centre and Association for Staff Cooperation with Developing Countries. http://www.impactalliance.org/ev02.php?ID=20281_201&ID2=DO_TOPIC. Accessed March 15, 2006.

Staatz, J. M., and C. K. Eicher. 1998. Agricultural development ideas in historical perspective. In *International agricultural development,* third edition, ed. C. K. Eicher and J. M. Staatz. Baltimore, Md., U.S.A.: Johns Hopkins University Press.

Stoneman, P. 1995. Introduction. In *Handbook of the economics of innovation and technological change,* ed. P. Stoneman. Oxford: Blackwell.

Teece, D. J., G. Pisano, and A. Shuen. 1997. Dynamic capabilities and strategic management. *Strategic Management Journal* 18 (7): 509–533.

Teubal, M. 1997. A catalytic and evolutionary approach to horizontal technology policies. *Research Policy* 25 (8): 1161–1188.

Thèves, J., L. Benedetto, and P. Larédo. 2007. Changing patterns of public research funding in France. *Science and Public Policy* 34 (6): 389–399.

Tidd, J., J. Bessant, and P. Pavitt. 1997. *Managing innovation. Integrating technological, market and organizational change.* Chichester, U.K.: John Wiley and Sons.

Toro, G., and N. Espinosa. 2003. *Los Fondos Competitivos para la Agricultura y el Desarrollo Rural: Fundamentos, aplicaciones y lecciones.* Costa Rica: Instituto Interamericano de Cooperación para la Agricultura.

Vandervert, L. R. 2003. The neurophysiological basis of innovation. In *The international handbook on innovation,* ed. L. V. Shavinina. London: Pergamon.

Vargas-Lundius, R. 2004. Remittances and rural development. Paper prepared for the Twenty-Seventh Session of IFAD's Governing Council, February 18–19. Rome: International Fund for Agricultural

Development. http://www.ifad.org/events/gc/27/roundtable/pl/discussion.pdf. Accessed May 4, 2009.

Vera-Cruz, A. O. 2004. *Cultura de la empresa, aprendizaje y capacidades tecnológicas. El Caso de las Cerveceras Mexicanas.* Mexico City: Universidad Autónoma Metropolitana, Asociación Mexicana de Directivos de la Investigación Aplicada y el Desarrollo Tecnológico, and Miguel Angel Porrúa.

Vera-Cruz, A. O., G. Dutrénit, J. M. Ekboir, G. Martínez V., and A. Torres Vargas. 2008. Virtues and limits of competitive funds to finance research and innovation in the agricultural sector: The case of the Mexican Produce Foundations. *Science and Public Policy* 35 (7): 501–513.

Verspagen, B. 2005. Innovation and economic growth. In *The Oxford handbook of innovation,* ed. J. Fagerberg, D. C. Mowery, and R. R. Nelson. Oxford and New York: Oxford University Press.

Von Krogh, G., K. Ichijo, and I. Nonaka. 2000. *Enabling knowledge creation. How to unlock the mystery of tacit knowledge and release the power of innovation.* New York: Oxford University Press.

Watts, D. J. 1999. *Small world. The dynamics of networks between order and randomness.* Princeton Studies in Complexity. Princeton, N.J., U.S.A.: Princeton University Press.

Weatherspoon, D. D., and T. Reardon. 2003. The rise of the supermarket in Africa: Implications for agrifood systems and the rural poor. *Development Policy Review* 21 (3): 333–355.

Weber, M. 1984. *Economía y sociedad.* Mexico City: Fondo de Cultura Económica.

Weisberg, R. W. 2003. Case studies of innovation: Ordinary thinking, extraordinary outcomes. In *The international handbook on innovation,* ed. L. V. Shavinina. London: Pergamon.

Welker, M. E., and A. R. Cox. 2006. A report on the research activities at research universities. *Research Management Review* 15 (1): 3–12.

World Bank. 2003. *Making services work for poor people.* World Development Report 2004. Washington, D.C.

———. 2005. *Pro-poor growth in the 1990s. Lessons and insights from 14 countries.* Washington, D.C.

———. 2006a. *Enhancing agricultural innovation: How to go beyond the strengthening of research systems.* Washington, D.C.

———. 2006b. *Institutional innovation in agricultural research and extension systems in Latin America and the Caribbean.* Washington, D.C.

———. 2007. *World development report 2008. Agriculture for development.* Washington, D.C.

Yin, R. K. 2003. *Case study research. Design and methods,* third edition. Applied Social Research Methods Series. Thousand Oaks, Calif., U.S.A.: Sage.

Yunez-Naude, A., and F. Barceinas Paredes. 2003. The agriculture of Mexico after ten years of NAFTA implementation. Paper elaborated for preparation of the report *NAFTA'S promise and reality: Lessons for the hemisphere.* Washington, D.C.: Carnegie Endowment for International Peace.

Zander, U., and B. Kogut. 1995. Knowledge and the speed of the transfer and imitation of organization capabilities: An empirical test. *Organization Science* 6 (1): 76–92.

Made in the USA
Monee, IL
08 November 2026

67984625R00085